CONTEMPORARY MATHEMATICS

Titles in this Series

Titles in this Series

ERRETT BISHOP:
Reflections on Him and His Research

Errett Bishop

CONTEMPORARY
MATHEMATICS

Volume 39

ERRETT BISHOP:
Reflections on Him
and His Research

Murray Rosenblatt, Editor

WITHDRAWN

AMERICAN MATHEMATICAL SOCIETY
Providence · Rhode Island

PROCEEDINGS OF THE MEMORIAL MEETING
FOR ERRETT BISHOP
HELD AT THE UNIVERSITY OF CALIFORNIA, SAN DIEGO
SEPTEMBER 24, 1983

These proceedings were prepared by the American Mathematical Society with partial support from the National Science Foundation Grant DMS 83-12106.

1980 *Mathematics Subject Classification*. Primary 03A05, 03E25, 03F65, 03E45, 36A03, 46J10, 46J20, 46J15, 32E10, 32C10.

Library of Congress Cataloging in Publication Data
Main entry under title:

Errett Bishop: reflections on him and his research.
    Bibliography: p.
    Contents: Vita of Errett Bishop—Publications of Errett Bishop—Schizophrenia in contem--porary mathematics/Errett Bishop—[etc.]
    1. Function algebras—Addresses, essays, lectures. 2. Mathematics—Philosophy—Addresses, essays, lectures. 3. Bishop, Errett, d. 1983—Addresses, essays, lectures. I. Rosenblatt, Murray. II. Bishop, Errett, d. 1983.
QA326,E75   1984   510   85-752
ISBN 0-8218-5040-7

214926

CONTENTS

PREFACE

Errett Bishop was distinguished for his mathematical work on function algebras and the foundations of mathematics. His research on function algebras and complex analysis was universally recognized. The research on foundations which came later was also remarkable. As is well known, many people have very strong opinions concerning foundations. The object here is to present a view of Errett Bishop as an individual, a colleague to many of us, and as a mathematician. A vita of Bishop is first given with a listing of his mathematical papers. This is followed by a paper of Bishop's titled "Schizophrenia in contemporary mathematics." This paper was distributed in conjunction with the Colloquium Lectures he delivered in 1973 at the Seventy-Eighth Meeting of the American Mathematical Society.

Bishop died April 14, 1983 at his home in La Jolla after a brief illness. The Mathematics Department of the University of California, San Diego, organized a meeting September 24, 1983 in his memory at which a number of his former colleagues were invited to speak about him and his work. They were Stefan Warschawski of UCSD, John Wermer of Brown University, John Kelley of UC Berkeley, Irving Glicksberg of the University of Washington, and Gabriel Stolzenberg of Northeastern University. Unfortunately Stolzenberg withdrew his paper on Errett Bishop. Other than that the papers are printed in full after Errett Bishop's paper. At the memorial meeting a scholarship in mathematics in Errett Bishop's name was established at UCSD, funded by a gift of his mother Mrs. Helen Bishop. To our sorrow, Irving Glicksberg died shortly after his manuscript of his paper was completed. Thanks are due to the Mathematics Department and the administration of UCSD for help in organizing the memorial meeting. A last paper on Bishop's work written by Metakides, Nerode and Shore concludes this selection of papers. This is preceded by a short paper of Nerode, Metakides and Constable on remembrances of Errett Bishop. Errett Bishop is survived by his wife Jane, his two sons, Edward and Thomas, and his daughter Rosemary.

La Jolla, California, 1984   Murray Rosenblatt

Some participants in the memorial meeting for Bishop. From left to right: I. Glicksberg, J. Wermer, S. Warschawski, H. Royden, J. Kelly, and M. Rosenblatt.

ASSOCIATION OF AUTHORS

Errett Bishop

Mathematics Department
University of California, San Diego

Robert Constable

Computer Science Department
Cornell University

Irving Glicksberg

Mathematics Department
University of Washington, Seattle

John Kelley

Mathematics Department
University of California, Berkeley

George Metakides

University of Patras

Anil Nerode

Mathematics Department
Cornell University

Halsey Royden

Mathematics Department
Stanford University

Richard Shore

Mathematics Department
Cornell University

Stefan Warschawski

Mathematics Department
University of California, San Diego

John Wermer

Mathematics Department
Brown University

VITA

Errett Bishop was born July 14, 1928. He received a Bachelor of Science degree in 1947, a Master of Science in 1949, and a Ph. D. in 1955 all from the University of Chicago. From 1950 to 1952 he was in the U.S. Army stationed at the Bureau of Standards in Washington, D.C. carrying out mathematical research in ordnance. His academic positions are listed:

Instructor of Mathematics   1954-1957

Assistant Professor of Mathematics   1957-1959

Associate Professor of Mathematics   1959-1962

Professor of Mathematics   1962-1965

all at the University of California, Berkeley.

Professor of Mathematics   1965-1983

University of California, San Diego.

Bishop held a Sloan Foundation Fellowship from 1958 to 1961. In 1961-1962 he was a member of the Institute for Advanced Study at Princeton. He was a Miller Fellow during the year 1964-1965. Bishop gave a number of distinguished invited lectures, among them an invited talk to the International Congress of Mathematicians at Moscow, USSR (1966), the Hedrick lectures of the Mathematical Association of America (1969), and the Colloquium Lectures of the American Mathematical Society (1973). He was elected a fellow of the American Academy of Arts and Sciences. His doctoral students at UCSD were James D. Brom (1974), Yuen-Kwok Chan (1969), Henry Cheng (1971), Merrill B. Goldberg (1969), John A. Nuber (1969) and Jonathan B. Tennenbaum (1973).

PUBLICATIONS OF ERRETT BISHOP

1. "Spectral theory for operators on a Banach space," Trans. Amer. Math. Soc., 86, No. 2 (November 1957), 414-445.

2. "Subalgebras of functions on a Riemann surface," Pacific J. Math., 8, No. 1 (1958).

3. "Measures orthogonal to polynomials," Proc. Nat'l. Acad. Sci., 44, No. 3 (March 1958), 278-280.

4. "The structure of certain measures," Duke Math. J., 25, No. 2 (June 1958), 283-290.

5. "Approximation by a polynomial and its derivatives on certain closed sets," Proc. Amer. Math. Soc., 9, No. 6 (December 1958), 946-953.

6. "A duality theorem for an arbitrary operator," Pacific J. Math., 9, No. 2 (1959).

7. "A minimal boundary for function algebras," Pacific J. Math., 9, No. 3 (1959).

8. "Some theorems concerning function algebras," Bull. Amer. Math. Soc., 65, No. 2 (March 1959), 77-78.

9. (with Karel de Leeuw) "The representations of linear functionals by measures on sets of extreme points," Annales de L'Institut Fourier, 9 (1959).

10. "Simultaneous approximation by a polynomial and its derivatives," Proc. Amer. Math. Soc., 10, No. 5 (October 1959), 741-743.

11. "Boundary measures of analytic differentials," Duke Math. J., 27, No. 3 (September 1960), 331-340.

12. "A generalization of the Stone-Weierstrass Theorem," Pacific J. Math., 11, No. 3 (1961).

13. (with R. R. Phelps) "A proof that every Banach space is subreflexive," Bull. Amer. Math. Soc., 67, No. 1 (January 1961), 97-98.

14. "Mappings of partially analytic spaces," Amer. J. Math., 83, No. 2 (April 1961), 209-242.

15. "Some global problems in the theory of functions of several complex variables," Amer. J. Math., 83, No. 3 (July 1961), 479-498.

16. "Partially analytic spaces," Amer. J. Math., 83, No. 4 (October 1961), 669-692.

17. "A general Rudin-Carleson Theorem," Proc. Amer. Math. Soc., 13, No. 1 (February 1962), 140-143.

18. "Analytic functions with values in a Frechet space," Pacific J. Math., 12, No. 4 (1962).

19. "Analyticity in certain Banach algebras," Trans. Amer. Math. Soc., 102, No. 3 (March 1962), 507-544.

20. (with R. R. Phelps) "The support functionals of a convex set," Proc. Symposia in Pure Math., VII (1963).

21. "Holomorphic completions, analytic continuations and the interpolation of semi-norms," Annals of Math., 78, No. 3 (November 1963), 468-500.

22. "Representing measures for points in a uniform algebra," Bull. Amer. Math. Soc., 70, No. 1 (January 1964), 121-122.

23. "Conditions for the analyticity of certain sets," Mich. Math. J., 11 (1964), 289-304.

24. "Uniform algebras," Proc. Conf. on Complex Analysis, Minneapolis (1964).

25. "Differentiable manifolds in complex Euclidean space," Duke Math. J., 32, No. 1 (March 1965), 1-22.

26. Constructive Methods in the Theory of Banach Algebras, in Function Algebras, 1966.

27. "An upcrossing inequality with applications," Mich. Math. J., 13 (1966), 1-13.

28. Foundations of Constructive Analysis, McGraw-Hill Book Company, 1967.

29. "A constructivization of abstract mathematical analysis," Proc. Inter. Congress Math., Moscow, 1966.

30. "Mathematics as a numerical language, intuitionism and proof theory," in Proc. Conf. Buffalo, New York, 1968, pp. 53-71. North Holland, Amsterdam, 1970.

31. (with Henry Cheng) "Constructive Measure Theory," to appear in Memoirs, Amer. Math. Soc.

32.   "The crisis in contemporary mathematics," Historia Mathematica, 2
       (1975), 507-517.

33.   Review of "Elementary Calculus," by H. Jerome Keisler, to appear in
       Amer. Math. Monthly.

34.   Schizophrenia in contemporary mathematics, lecture.

Contemporary Mathematics
Volume 39, 1985

SCHIZOPHRENIA IN CONTEMPORARY MATHEMATICS

Errett A. Bishop

During the past ten years I have given a number of lectures on the subject of constructive mathematics. My general impression is that I have failed
to communicate a real feeling for the philosophical issues involved. Since I
am here today, I still have hopes of being able to do so. Part of the difficulty is the fear of seeming to be too negativistic and generating too much
hostility. Constructivism is a reaction to certain alleged abuses of classical
mathematics. Unpalatable as it may be to have those abuses examined, there
is no other way to understand the motivations of the constructivists.

Brouwer's criticisms of classical mathematics were concerned with
what I shall refer to as "the debasement of meaning." His incisive criticisms
were one of his two main contributions to constructivism. (His other was to
establish a new terminology, involving a re-interpretation of the usual connectives and quantifiers, which permits the expression of certain important
distinctions of meaning which the classical terminology does not.)

The debasement of meaning is just one of the trouble spots in contemporary mathematics. Taken all together, these trouble spots indicate that
something is lacking, that there is a philosophical deficit of major propor-
tions. What it is that is lacking is perhaps not clear, but the lack in all of its
aspects constitutes a syndrome I shall tentatively describe as "schizophrenia."

One could probably make a long list of schizophrenic attributes of contemporary mathematics, but I think the following short list covers most of the
ground: rejection of common sense in favor of formalism; debasement of
meaning by wilful refusal to accommodate certain aspects of reality; inappro-
priateness of means to ends; the esoteric quality of the communication; and
fragmentation.

Common sense is a quality that is constantly under attack. It tends to be supplanted by methodology, shading into dogma. The codification of insight is commendable only to the extent that the resulting methodology is not elevated to dogma and thereby allowed to impede the formation of new insight. Contemporary mathematics has witnessed the triumph of formalist dogma, which had its inception in the important insight that most arguments of modern mathematics can be broken down and presented as successive applications of a few basic schemes. The experts now routinely equate the panorama of mathematics with the productions of this or that formal system. Proofs are thought of as manipulations of strings of symbols. Mathematical philosophy consists of the creation, comparison, and investigation of formal systems. Consistency is the goal. In consequence meaning is debased, and even ceases to exist at a primary level.

The debasement of meaning has yet another source, the wilful refusal of the contemporary mathematician to examine the content of certain of his terms, such as the phrase "there exists." He refuses to distinguish among the different meanings that might be ascribed to this phrase. Moreover he is vague about what meaning it has for him. When pressed he is apt to take refuge in formalistics, declaring that the meaning of the phrase and the statements of which it forms a part can only be understood in the context of the entire set of assumptions and techniques at his command. Thus he inverts the natural order, which would be to develop meaning first, and then to base his assumptions and techniques on the rock of meaning. Concern about this debasement of meaning is a principal force behind constructivism.

Since meaning is debased and common sense is rejected, it is not surprising to find that the means are inappropriate to the ends. Applied mathematics makes much of the concept of a model, as a tool for dealing with reality by mathematical means. When the model is not an adequate representation of reality, as happens only too often, the means are inappropriate. One gets the impression that some of the model-builders are no longer interested in reality. Their models have become autonomous. This has clearly happened in mathematical philosophy: the models (formal systems) are accepted as the preferred tools for investigation the nature of mathematics, and even as the font of meaning.

Everyone who has taught undergraduate mathematics must have been impressed by the esoteric quality of the communication. It is not natural for "A implies B" to mean "not A or B, " and students will tell you so if you give them the chance. Of course, this is not a fatal objection. The question is, whether the standard definition of implication is useful, not whether it is natural. The constructivist, following Brouwer, contends that a more natural definition of implication would be more useful. This point will be developed later. One of the hardest concepts to communicate to the undergraduate is the concept of a proof. With good reason -- the concept is esoteric. Most mathematicians, when pressed to say what they mean by a proof, will have recourse to formal criteria. The constructive notion of proof by contrast is very simple, as we shall see in due course. Equally esoteric, and perhaps more troublesome, is the concept of existence. Some of the problems associated with this concept have already been mentioned, and we shall return to the subject again. Finally, I wish to point out the esoteric nature of the classical concept of truth. As we shall see later, truth is not a source of trouble to the constructivist, because of his emphasis on meaning.

The fragmentation of mathematics is due in part to the vastness of the subject, but it is aggravated by our educational system. A graduate student in pure mathematics may or may not be required to broaden himself by passing examinations in various branches of pure mathematics, but he will almost certainly not be required or even encouraged to acquaint himself with the philosophy of mathematics, its history, or its applications. We have geared ourselves to producing research mathematicians who will begin to write papers as soon as possible. This anti-social and anti-intellectual process defeats even its own narrow ends. The situation is not likely to change until we modify our conception of what mathematics is. Before important changes will come about in our methods of education and our professional values, we shall have to discover the significance of theorem and proof. If we continue to focus attention on the process of producing theorems, and continue to devalue their content, fragmentation is inevitable.

By devaluation of content I mean the following. To some pure mathematicians the only reason for attaching any interpretation whatever to theorem and proof is that the process of producing theorems and proofs is thereby facilitated. For them content is a means rather than the end. Others feel that it is important to have some content, but don't especially care to find out what it is. Still others, for whom Gödel (see for example [16]) seems to be a leading spokesman, do their best to develop content within the accepted framework of platonic idealism. One suspects that the majority of pure mathematicians, who belong to the union of the first two groups, ignore as much content as they possibly can. If this suspicion seems unjust, pause to consider the modern theory of probability. The probability of an event is commonly taken to be a real number between 0 and 1. One might naively expect that the probabilists would concern themselves with the computation of such real numbers. If so, a quick look at any one of a number of modern texts, for instance the excellent book of Doob [14], should suffice to disabuse him of that expectation. Fragmentation ensues, because much if not most of the theory is useless to someone who is concerned with actually finding probabilities. He will either develop his own semi-independent theories, or else work with ad hoc techniques and rules of thumb. I do not claim that reinvolvement of the probabilists with the basic questions of meaning would of itself reverse the process of fragmentation of their discipline, only that it is a necessary first step. In recent years a small number of constructivists (see [3], [9], [10], [11], [12], [23], and [24]) have been trying to help the probabilists take that step. Whether their efforts will ultimately be appreciated remains to be seen.

When I attempt to express in positive terms that quality in which contemporary mathematics is deficient, the absence of which I have characterized as "schizophrenia," I keep coming back to the term "integrity." Not the integrity of an isolated formalism that prides itself on the maintenance of its own standards of excellence, but an integrity that seeks common ground in the researches of pure mathematics, applied mathematics, and such mathematically oriented disciplines as physics; that seeks to extract the maximum meaning from each new development; that is guided primarily by considerations of content rather than elegance and formal attractiveness; that sees to it that the mathematical representation of reality does not degenerate into a game; that seeks to understand the place of mathematics in contemporary society. This integrity may not be possible of realization, but that is not important. I like to think of constructivism as one attempt to realize at least certain aspects of this idealized integrity. This presumption at least has the possible merit of preventing constructivism from becoming another game, as some constructivisms have tended to do in the past.

In discussing the principles of constructivism, I shall try to separate those aspects of constructivism that are basic to the philosophy from those that are merely convenient (or inconvenient, as the case may be). Four principles stand out as basis:

(A) Mathematics is common sense.

(B) Do not ask whether a statement is true until you know what it means.

(C) A proof is any completely convincing argument.

(D) Meaningful distinctions deserve to be maintained.

Surprisingly many brilliant people refuse to apply common sense to mathematics. A frequent attitude is that the formalization of mathematics has been of great value, because the formalism constitutes a court of last resort to settle any disputes that might arise concerning the correctness of a proof. Common sense tells us, on the contrary, that if a proof is so involved that we are unable to determine its correctness by informal methods, then we shall not be able to test it by formal means either. Moreover the formalism cannot be used to settle philosophical disputes, because the formalism merely reflects the basic philosophy, and consequently philosophical disagreements are bound to result in disagreements about the validity of the formalism.

Principle (B) resolves the problem of constructive truth. For that matter, it would resolve the problem of classical truth if the classical mathematicians would accept it. We might say that truth is a matter of convention. This simply means that all arguments concerning the truth or falsity of any given statement about which both parties possess the same relevant facts occur because they have not reached a clear agreement as to what the statement means. For instance in response to the inquiry "Is it true the constructivists believe that not every bounded monotone sequence of real numbers converges?," if I am tired I answer "yes." Otherwise I tell the questioner that my answer will depend on what meaning he wishes to assign to the statement (*), that every bounded monotone sequence or real numbers converges. Moreover I tell him that once he has assigned a precise meaning to statement (*), then my answer to his question will probably be clear to him before I give it. The two meanings commonly assigned to (*) are the classical and the constructive. It seems to me that the classical mathematician is not as precise as he might be about the meaning he assigns to such a statement. I shall show you later one simple and attractive approach to the problem of meaning in classical mathematics. However in the case before us the intuition at least is

clear. We represent the terms of the sequence by vertical marks marching
to the right, but remaining to the left of the bound B.

$$| \; | \; | \; | \; | \; | \; | \; | \; |||| \; \cdots \qquad |$$
$$B$$

The classical intuition is that the sequence gets cramped, because there are
infinitely many terms, but only a finite amount of space available to the left
of B. Thus it has to pile up somewhere. That somewhere is its limit L.

$$| \; | \; | \; | \; | \; | \; \cdots \qquad | \qquad |$$
$$L \qquad B$$

The constructivist grants that some sequences behave in precisely this way.
I call those sequences stupid. Let me tell you what a smart sequence will do.
It will pretend to be stupid, piling up at a limit (in reality a false limit) L_f.
Then when you have been convinced that it really is piling up at L_f, it will
take a jump and land somewhere to the right!

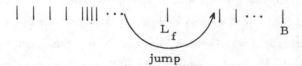

jump

Let us postpone a serious discussion of this example until we have discussed
the constructive real number system. The point I wish to make now is that
under neither interpretation will there be any disagreement as to the truth of
(*), once that interpretation has been fixed and made precise.

Whenever a student asks me whether a proof he has given is correct,
before answering his question I try to discover his concept of what constitutes
a proof. Then I tell him my own concept, (C) above, and ask him whether he
finds his argument completely convincing, and whether he thinks he has expressed himself clearly enough so that other informed and intelligent people
will also be completely convinced.

Clearly it is impossible to accept (C) without accepting (B), because it
doesn't make sense to be convinced that something is true unless you know
what it means.

The question often arises, whether a constructivist would accept a non-
constructive proof of a numerical result involving no existential quantifiers,
such as Goldbach's conjecture or Fermat's last theorem. My answer is supplied by (C): I would want to examine the proof to see whether I found it completely convincing. Perhaps one should keep an open mind, but I find it hard
to believe that I would find any proof that relied on the principle of the

excluded middle for instance completely convincing. Fortunately the problem
is hypothetical, because such proofs do not seem to arise. It does raise the
interesting point that a classically acceptable proof of Goldbach's conjecture
might not be constructively acceptable, and therefore the classical and the
constructive interpretations of Goldbach's conjecture must differ in some
fundamental respect. We shall see later that this is indeed the case.

Classical mathematics fails to observe meaningful distinctions having to
do with integers. This basic failure reflects itself at all levels of the classical development of mathematics. Consider the number n_0, defined to be 0 if
the Riemann hypothesis is true and 1 if it is false. The constructivist does
not wish to prevent the classicist from working with such numbers (although
he may personally believe that their interest is limited). He does want the
classicist to distinguish such numbers from numbers which can be "computed," such as the number n_1 of primes less than $10^{10^{10}}$. Classical
mathematicians do concern themselves sporadically with whether numbers
can be "computed," but only on an ad hoc basis. The distinction is not observed in the systematic development of classical mathematics, nor would the
tools available to the classicist permit him to observe the distinction systematically even if he were so inclined.

The constructivists are frequently accused of displaying the same insen-
sitivity to shades of meaning of which they accuse the classicist, because they
do not distinguish between numbers that can be computed in principle, such as
the number n_1 defined above, and numbers that can be computed in fact.
Thus they violate their own principle (D). This is a serious accusation, and
one that is not easy to refute. Rather than attempting to refute it, I shall give
you my personal point of view. First, it may be demanding too much of the
constructivists to ask them to lead the way in the development of usable and
systematic methods for distinguishing computability in principle from computability in fact. If and when such methods are found, the constructivists will
gratefully incorporate them into their mathematics. Second, it is by no means
clear that such methods are going to be found. There is no fast distinction between computability in principle and in fact, because of the constant progress
of the state of the art among other reasons. The most we can hope for is
some good systematic measure of the efficiency of a computation. Until such
is found, the problem will continue to be treated on an ad hoc basis.

I was careful not to call the number n_0 defined above an integer.
Whether we do call it an integer is of no real importance, as long as we distinguish it in some way from numbers such as n_1. For instance we might
call n_0 an integer and call n_1 a constructive integer. The constructivists
have not accepted this terminology, in part because of Brouwer's influence,
but also because it does not accord with their estimate of the relative importance of the two concepts. I shall reserve the term "integer" for what a
classicist might call a constructive integer, and put aside, at least for now,
the problem of what would be an appropriate term for what is classically
called an integer (assuming that the classical notion of an integer is indeed
viable).

Thus we come to the crucial question, "What is an integer?" As we
have already seen, the question is badly phrased. We are really looking for
a definition of an integer that will be an efficient tool for developing the full
content of mathematics. Since it is clear that we always work with representations of integers, rather than integers themselves (whatever those may
be), we are really trying to define what we mean by a representation of an
integer. Again, an integer is represented only when some intelligent agent
constructs the representation, or establishes the convention that some artifact
constitutes a representation. Thus in its final version the question is, "How
does one represent an integer?" In practice we shall not be so meticulous as
all this in our use of language. We shall simply speak of integers, with the
understanding that we are really speaking of their representations. This
causes no harm, because the original concept of an integer, as something
invariant standing behind all of its representations, has just been seen to be
superfluous. Moreover we shall not constantly trouble to point out that
(representations of) integers exist only by virtue of conventions established by
groups of intelligent beings. After this preliminary chatter, which may seem
to have been unnecessary, we present our definition of an integer, dignified by
the title of the

Fundamental Constructivist Thesis

Every integer can be converted in principle to decimal form by a finite,
purely routine, process.

Note the phrase "in principle." It means that although we should be
able to program a computer to produce the decimal form of any given integer,

there are cases in which it would be naive to run the program and wait around for the result.

Everything else about integers follows from the above thesis plus the rules of decimal arithmetic that we learned in elementary school. Two integers are equal if their decimal representations are equal in the usual sense. The order relations and the arithmetic of integers are defined in terms of their decimal representations.

With the constructive definition of the integers, we have begun our study of the technical implementation of the constructivist philosophy. Our point of view is to describe the mathematical operations that can be carried out by finite beings, man's mathematics for short. In contrast, classical mathematics concerns itself with operations that can be carried out by God. For instance, the above number n_0 is classically a well-defined integer because God can perform the infinite search that will determine whether the Riemann hypothesis is true. As another example, the smart sequences previously discussed may be able to outwit you and me (or any other finite being), but they will not be able to outwit God. That is why statement (*) is true classically but not constructively.

You may think that I am making a joke, or attempting to put down classical mathematics, by bringing God into the discussion. This is not true. I am doing my best to develop a secure philosophical foundation, based on meaning rather than formalistics, for current classical practice. The most solid foundation available at present seems to me to involve the consideration of a being with non-finite-powers -- call him God or whatever you will -- in addition to the powers possessed by finite beings.

What powers should we ascribe to God? At the very least, we should credit him with limited omniscence, as described in the following limited principle of omniscence (LPO for short): If $\{n_k\}$ is any sequence of integers, then either $n_k = 0$ for all k or there exists a k with $n_k \neq 0$. By accepting LPO as valid, we are saying that the being whose capabilities our mathematics describes is able to search through a sequence of integers to determine whether they all vanish or not.

Let us return to the technical development of constructive mathematics, since it is simpler, and postpone the further consideration of classical mathematics until later. Our first task is to develop an appropriate language to describe the mathematics of finite beings. For this we are indebted to

Brouwer. (See references [1], [6], [15], [20], and [21] for a more complete exposition that we are able to give here.) Brouwer remarked that the
meanings customarily assigned to the terms "and," "or," "not," "implies,"
"there exists," and "for all" are not entirely appropriate to the constructive
point of view, and he introduced more appropriate meanings as necessary.

The connective "and" causes no trouble. To prove "A and B," we must
prove A and also prove B, as in classical mathematics. To prove "A or B"
we must give a finite, purely routine method which after a finite number of
steps either leads to a proof of A or to a proof of B. This is very different
from the classical use of "or"; for example, LPO is true classically, but we
are not entitled to assert it constructively because of the constructive meaning
of "or."

The connective "implies" is defined classically by taking "A implies B"
to mean "not A or B." This definition would not be of much value constructively. Brouwer therefore defined "A implies B" to mean that there exists an
argument which shows how to convert an arbitrary proof of A into a proof of
B. To take an example, it is clear that "{(A implies B) and (B implies C)}
implies (A implies C)" is always true constructively; the argument that con-
verts arbitrary proofs of "A implies B" and "B implies C" into a proof of
"A implies C" is the following: given any proof of A, convert it into a proof
of C by first converting it into a proof of B and then converting that proof
into a proof of C.

We define "not A" to mean that A is contradictory. By this we mean
that it is inconceivable that a proof of A will ever be given. For example,
"not 0 = 1" is a true statement. The statement "0 = 1" means that when the
numbers "0" and "1" are expressed in decimal form, a mechanical comparison of the usual sort checks that they are the same. Since they are already in
decimal form, and the comparison in question shows they are not the same, it
is impossible by correct methods to prove that they are the same. Any such
proof would be defective, either technically or conceptually. As another example, "not (A and not A)" is always a true statement, because if we prove
not A it is impossible to prove A -- therefore, it is impossible to prove
both.

Having changed the meaning of the connectives, we should not be surprised to find that certain classically accepted modes of inference are no longer correct. The most important of these is the principle of the excluded

middle -- "A or not A." Constructively, this principle would mean that we had a method which, in finitely many purely routine steps, would lead to a proof of disproof of an arbitrary mathematical assertion A. Of course we have no such method, and nobody has the least hope that we ever shall. It is the principle of the excluded middle that accounts for almost all of the important unconstructivities of classical mathematics. Another incorrect principle is "(not not A) implies A." In other words, a demonstration of the impossibility of the impossibility of a certain construction, for instance, does not constitute a method for carrying out that construction.

I could proceed to list a more or less complete set of constructively valid rules of inference involving the connectives just introduced. This would be superfluous. Now that their meanings have been established, the rest is common sense. As an exercise, show that the statement

$$\text{"}(A \longrightarrow 0 = 1) \longleftrightarrow \text{not } A\text{"}$$

is constructively valid.

The classical concept of a set as a collection of objects from some pre-existent universe is clearly inappropriate constructively. Constructive mathematics does not postulate a pre-existent universe, with objects lying around waiting to be collected and grouped into sets, like shells on a beach. The entities of constructive mathematics are called into being by the constructing intelligence. From this point of view, the very question "What is a set?" is suspect. Rather we should ask the question, "What must one do to construct a set?" When the question is posed this way, the answer is not hard to find.

Definition. To construct a set, one must specify what must be done to construct an arbitrary element of the set, and what must be done to prove two arbitrary elements of the set are equal. Equality so defined must be shown to be an equivalence relation.

As an example, let us construct the set of rational numbers. To construct a rational number, define integers p and q and prove that $q \neq 0$. To prove that the rational numbers p/q and p_1/q_1 are equal, prove $pq_1 = p_1 q$.

While we are on the subject, we might as well define a function $f: A \to B$. It is a rule that to each element x of A associates an element $f(x)$ of B, equal elements of B being associated to equal elements of A.

The notion of a subset A_0 of a set A is also of interest. To construct an element of A_0, one must first construct an element of A, and then prove that the element so constructed satisfies certain additional conditions, characteristic of the particular subset A_0. Two elements of A_0 are equal if they are equal as elements of A.

Contrary to classical usage, the scope of the equality relation never extends beyond a particular set. Thus it does not make sense to speak of elements of different sets as being equal, unless possibly those different sets are both subsets of the same set. This is because for the constructivist equality is a convention, whose scope is always a given set; all this is conceptually quite distinct from the classical concept of equality as identity. You see now why the constructivist is not forced to resort to the artifice of equivalence classes!

After this long digression, consider again the quantifiers. Let $A(x)$ be a mathematical assertion depending on a parameter x ranging over a set S. To prove "$\forall xA(x)$," we must give a method which to each element x of S associates a proof of $A(x)$. Thus the meaning of the universal quantifier "$\forall$" is essentially the same as it is classically.

We expect the existential quantifier "$\exists$," on the other hand, to have a new meaning. It is not clear to the constructivist what the classicist means when he says "there exists." Moreover, the existential quantifier is just a glorified version of "or," and we know that a reinterpretation of this connective was necessary. Let the variable x range over the set S. Then to prove "$\exists xA(x)$" we must construct an element x_0 of S, according to the principles laid down in the definition of S, and then prove the statement "$A(x_0)$."

Again, certain classical uses of the quantifiers fail constructively. For example, it is not correct to say that "not $\forall xA(x)$ implies $\exists x$ not $A(x)$." On the other hand, the implication "not $\exists xA(x)$ implies $\forall x$ not $A(x)$" is constructively valid. I hope all this accords with your common sense, as it does with mine.

Perhaps you see an objection to these developments -- that they appear to violate constructivist principle (D) above. By accommodating our terminology to the mathematics of finite beings, have we not replaced the classical system, that does not permit the systematic development of constructive meaning, by a system that does not permit the systematic development of classical meaning? In my opinion the exact opposite is true -- the constructive terminology just introduced affords as good a framework as is presently available for expressing the content of <u>classical</u> mathematics.

If you wish to do classical mathematics, first decide what non-finite attributes you are willing to grant to God. You may wish to grant him LPO and no others. Or you may wish to be more generous and grant him EM, the principle of the excluded middle, possibly augmented by some version of the axiom of choice. When you have made your decision, avail yourself of all the apparatus of the constructivist, and augment it by those additional powers (LPO or EM or whatever) that you have granted to God. Although you will be able to prove more theorems than the constructivist will, because your being is more powerful than his, his theorems will be more meaningful than yours. Moreover to each of your theorems he will be able to associate one of his, having exactly the same meaning. For example, if LPO is the only non-finite attribute of your God, then each of your theorems "A" he will restate and prove as "LPO implies A." Clearly the meaning will be preserved. On the other hand, if he proves a theorem "B," you will also be able to prove "B," but your "B" will be less meaningful than his. The classical interpretation of even such simple results as Goldbach's conjecture is weaker than the constructive interpretation. In both cases the same phenomena -- the results of certain finitely performable computations -- are predicted, but the degree of conviction that the predicted phenomena will actually be observed is greater in the constructive case, because to trust the classical predictions one must believe in the theoretical validity of the concept of a God having the specified attributes, whereas to trust the constructive predictions one must only believe in the theoretical validity of the concept of a being who is able to perform arbitrarily involved finite operations.

It would thus appear that even a constructive proof of such a result as "the number of zeros in the first n digits of the decimal expansion of π does not exceed twice the number of ones" would leave us in some doubt as to whether the prediction is correct for any particular value of n, say for n = 1000. We have brought mathematics down to the gut level. My gut tells me to trust the constructive prediction and be wary of the classical prediction. I see no reason that yours should not tell you to trust both, or to trust neither.

In common with other constructivists, I also have gut feelings about the relative merits of the classical and constructive versions of those results which, unlike Goldbach's conjecture, assert the existence of certain quantities.

If we let "A" be any such result, with the constructive interpretation, then the constructive version of the corresponding classical result will be (for instance) "LPO → A," as we have seen. My feeling is that is likely to be worth whatever extra effort it takes to prove "A" rather than "LPO → A."

The linguistic developments I have outlined could be taken as the basis for a formalization of constructive (and therefore of classical) mathematics. So as not to create the wrong impression, I wish to emphasize again certain points that have already been made.

Formalism

> The devil is very neat. It is his pride
> To keep his house in order. Every bit
> Of trivia has its place. He takes great pains
> To see that nothing ever does not fit.
>
> And yet his guests are queasy. All their food,
> Served with a flair and pleasant to the eye,
> Goes through like sawdust. Pity the perfect host!
> The devil thinks and thinks and he cannot cry.

Constructivism

> Computation is the heart
> Of everything we prove.
> Not for us the velvet wisdom
> Of a softer love.
>
> If Aphrodite spends the night,
> Let Pallas spend the day.
> When the sun dispels the stars
> Put your dreams away.

There are at least two reasons for developing formal systems for constructive mathematics. First, it is good to state as concisely and systematically as we are able some of the objects, constructions, terminology, and methods of proof. The development of formal systems that catch these aspects of constructive practice should help to sharpen our understanding of how best to organize and communicate the subject. Second and more important, informal mathematics is the appropriate language for communicating with people, but formal mathematics is more appropriate for communicating with machines. Modern computer languages (see the report [30], for example), while rich in facilities, seem to be lacking in philosophical scope. It might be worthwhile to investigate the possibility that constructive mathematics would

afford a solid philosophical basis for the theory of computation, and constructive formalism a point of departure for the development of a better computer
language. Certainly recursive function theory, which has played a central
role in the philosophy of computation, is inadequate to the task.

The development of a constructive formalism at any given level would
seem to be no more difficult than the development of a classical formalism at
the same level. See [17], [18], [20], [21], [22], and [27] for examples.
For a discussion of constructive formalism as a computer language, see [2].

Let us return to the technical development of constructive mathematics,
and ask what is meant constructively by a function $f: \mathbb{Z} \to \mathbb{Z}$ (where $\mathbb{Z}$ is the
set of integers). We improve the classical treatment right away - instead of
talking about ordered pairs, we talk about rules. Our definition takes a function $F: \mathbb{Z} \to \mathbb{Z}$ to be a rule that associates to each (constructively defined)
integer n a (constructively defined) integer $f(n)$, equal values being associated to equal arguments. For a given argument n, the requirement that $f(n)$
be constructively defined means that its decimal form can be computed by a
finite, purely routine process. That's all there is to it. Functions $f: \mathbb{Z} \to \mathbb{Q}$,
$f: \mathbb{Q} \to \mathbb{Q}$, $f: \mathbb{Z}^+ \to \mathbb{Q}$ are defined similarly. (Here $\mathbb{Q}$ is the set of rational
numbers and $\mathbb{Z}^+$ the set of positive integers.) A function with domain $\mathbb{Z}^+$ is
called a sequence, as usual.

Now that we know what a sequence of rational numbers is, it is easy to
define a real number. A real number is a Cauchy sequence of rational numbers! Again, I have improved on the classical treatment, by not mentioning
equivalence classes. I shall never mention equivalence classes. To be sure
we completely understood this definition, let us expand it a bit. Real numbers
are not pre-existent entities, waiting to be discovered. They must be constructed. Thus it is better to describe how to construct a real number, than
to say what it is. To construct a real number, one must

  (a) construct a sequence $\{x_n\}$ of rational numbers,
  (b) construct a sequence $\{N_n\}$ of integers,
  (c) prove that for each positive integer n we have

$$|x_i - x_j| \le \frac{1}{n} \quad \text{whenever} \quad i \ge N_n \quad \text{and} \quad j \ge N_n.$$

Of course, the proof (c) must be constructive, as well as the constructions (a)
and (b).

Two real numbers $\{a_n\}$ and $\{b_n\}$ (the corresponding convergence parameters (b) and proofs (c) are assumed without explicit mention) are said to be <u>equal</u> if for each positive integer k there exists a positive integer N_k such that $|a_n - b_n| \le \frac{1}{k}$ whenever $n \ge N_k$. It can be shown that this notion of equality is an equivalence relation. Addition and multiplication of real numbers are also defined in the same way as they are defined classically. The order relation, on the other hand, is more interesting. If $a = \{a_n\}$ and $b = \{b_n\}$ are real numbers, we define $a < b$ to mean that there exist positive integers M and N such that $a_n \le b_n - \frac{1}{M}$ whenever $n \ge N$. Then it is easily shown that $a < b$ and $b < c$ imply $a < c$, that $a < b$ implies $a - c < b - c$, and so forth. Some care must be exercised in defining the relation $\le$. We could define $a \le b$ to mean that either $a < b$ or $a = b$. An alternate definition would be to define it to mean that $b < a$ is contradictory. We shall not use either of these, although our definition turns out to be equivalent to the latter.

DEFINITION. $a \le b$ means that for each positive integer M there exists a positive integer N such that $b_n \ge a_n - \frac{1}{M}$ whenever $n \ge N$.

To make the choice of this definition plausible, I shall construct a certain real number H.

$$H = \sum_{n=1}^{\infty} \alpha_n 2^{-n}$$

where $\alpha_n = 0$ in case every even integer between 4 and n is the sum of two primes, and $\alpha_n = 1$ otherwise. (More precisely, H is given by the Cauchy sequence $\{a_n\}$, with

$$a_n = \sum_{k=1}^{n} \alpha_k 2^{-k} \quad,$$

and the sequence $\{N_n\}$ of convergence parameters, where $N_n = n$.) Clearly we wish to have $H \ge 0$. It certainly is according to the definition we have chosen. (The real number 0 of course is the Cauchy sequence of rational numbers all of whose terms are 0.) On the other hand, we would not be entitled to assert that $H \ge 0$ if we had defined $H \ge 0$ to mean that either $H > 0$ or $H = 0$, because the assertion "$H > 0$ or $H = 0$" means that we have a finite, purely routine method for deciding which; in this case, a finite, purely routine method for proving or disproving Goldbach's conjecture!

Most of the usual theorems about $\leq$ and $<$ remain true constructively, with the exception of trichotomy. Not only does the usual form "a < b or a = b or a > b" fail, but such weaker forms as "a < b or a $\geq$ b," or even "a $\leq$ b or a $\geq$ b" fail as well. For example, we are not entitled to assert "0 < H or 0 = H or 0 > H." If we consider the closely related number

$$H' = \sum_{n=1}^{\infty} \alpha_{2n} (-2)^{-n} ,$$

we are not even entitled to assert that "H' $\geq$ 0 or H' $\leq$ 0."

Since trichotomy is so fundamental, we might expect constructive mathematics to be hopelessly enfeebled because of its failure. The situation is saved, because trichotomy does have a constructive version, which of course is considerably weaker than the classical.

THEOREM. For arbitrary real numbers a, b, and c, with a < b, either c > a or c < b .

PROOF. Choose integers M and N_0 such that $a_n \leq b_n - \frac{1}{M}$ whenever $n \geq N_0$. Choose integers N_a, N_b, and N_c such that $|a_n - a_m| \leq (6M)^{-1}$ whenever $n, m \geq N_a$, $|b_n - b_m| \leq (6M)^{-1}$ whenever $n, m \geq N_b$, $|c_n - c_m| \leq (6M)^{-1}$ whenever $n, m \geq N_c$. Set $N = \max\{N_0, N_a, N_b, N_c\}$. Since a_N, b_N, and c_N are all rational numbers, either

$$c_N < \frac{1}{2} (a_N + b_N) \quad \text{or} \quad c_N \geq \frac{1}{2} (a_N + b_N) .$$

. Consider first the case $c_N \geq \frac{1}{2} (a_N + b_N)$. Since $a_N \leq b_N - M^{-1}$, it follows that $a_N \leq c_N - (2M)^{-1}$. For each $n \geq N$ we therefore have

$$a_n \leq a_N + (6M)^{-1} \leq c_N - (2M)^{-1} + (6M)^{-1}$$

$$\leq c_n + (6M)^{-1} - (2M)^{-1} + (6M)^{-1} = c_n - (6M)^{-1} .$$

Therefore, a < c. In the other case, $c_N < \frac{1}{2} (a_N + b_N)$, it follows similarly that c < b. This completes the proof of the theorem.

Do not be deceived by the use of the word "choose" in the above proof, which is simply a carry-over from classical usage. No choice is involved,

because M and N_0, for instance, are fixed positive integers, defined explicitly by the proof of the inequality $a < b$. Of course we could decide to substi-
tute other values for the original values of M and N_0, if we desired, so
some choice is possible should we wish to exercise it. If we do not explicitly
state what choice we wish to make, it will be assumed that the values of M
and N_0 given by the proof of $a < b$ are chosen.

The number H, which is constructively a well-defined real number, is
classically rational, because if the Goldbach conjecture is true then $H = 0$,
and if the conjecture is false then $H = 2^{-n+1}$, where n is the first even
integer for which it fails. We are not entitled to assert constructively that H
is rational: if it is rational, then either $H = 0$ or $H \neq 0$, meaning that either
Goldbach's conjecture is true or else it is false; and we are not entitled to
assert this constructively, until we have a method for deciding which. We are
not entitled to assert H is irrational either, because if H is irrational, then
$H \neq 0$, therefore Goldbach's conjecture is false, therefore H is the rational
number 2^{-n+1}, a contradiction! Thus H cannot be asserted to be rational,
although its irrationality is contradictory. (I am indebted to Halsey Royden
for this amusing observation.)

It is easy to prove the existence of many irrational numbers, by proving
the uncountability of the real numbers, as a corollary of the Baire category
theorem. For the present, let us merely remark that $\sqrt{2}$ is irrational. Of
course, $\sqrt{2}$ can be defined by constructing successive decimal approximations. It is therefore constructively well-defined. The classical proof of the
irrationality of $\sqrt{2}$ shows that if p/q is any rational number then $p^2/q^2 \neq 2$.
Since both p^2/q^2 and 2 can be written with denominator q^2, it follows that

$$\left| \frac{p}{q} - \sqrt{2} \right| \cdot \left| \frac{p}{q} + \sqrt{2} \right| = \left| \frac{p^2}{q^2} - 2 \right| \geq \frac{1}{q^2} .$$

Since clearly $p/q \neq \sqrt{2}$ in case $p/q < 0$ or $p/q > 2$, to show that $p/q \neq \sqrt{2}$
we may assume $0 \leq p/q \leq 2$. Then

$$\left| \frac{p}{q} - \sqrt{2} \right| \geq \left| \frac{p}{q} + \sqrt{2} \right|^{-1} \cdot \frac{1}{q^2} \geq |2 + 2|^{-1} \cdot \frac{1}{q^2} = \frac{1}{4q^2} .$$

Therefore, $\sqrt{2} \neq p/q$. Thus $\sqrt{2}$ is (constructively) irrational.

The failure of the usual form of trichotomy means that we must be careful in defining absolute values and maxima and minima of real numbers. For

example, if $x = \{x_n\}$ is a real number, with sequence $\{N_n\}$ of convergence parameters, then $|x|$ is defined to be the Cauchy sequence $\{|x_n|\}$ of rational numbers (with sequence $\{N_n\}$ of convergence parameters). Similarly, $\min\{x, y\}$ is defined to be the Cauchy sequence $\{\min\{x_n, y_n\}\}_{n=1}^{\infty}$, and $\max\{x, y\}$ to be $\{\max\{x_n, y_n\}\}_{n=1}^{\infty}$.

This definition of $\min$, in particular, has an amusing consequence. Consider the equation

$$x^2 - xH' = 0 .$$

Clearly 0 and the number H' are solutions. Are they the only solutions? It depends on what we mean by "only." Clearly $\min\{0, H'\}$ is a solution, and we are unable to identify it with either 0 or H'. Thus it is a third solution! The reader might like to amuse himself looking for others. This discussion incidentally makes the point that if the product of two real numbers is 0 we are not entitled to conclude that one of them is 0. (For example, $x(x - H') = 0$ does not imply that $x = 0$ or $x - H' = 0$: set $x = \min\{0, H'\}$.)

The constructive real number system as I have described it is not accepted by all constructivists. The intuitionists and the recursive function theorists have other versions.

For Brouwer, and his followers (the intuitionists), the constructive real numbers described above do not constitute all of the real number system. In addition there are incompletely determined real numbers, corresponding to sequences of rational numbers whose terms are not specified by a master algorithm. Such sequences are called "free-choice sequences," because the creating subject, who defines the sequence, does not completely commit himself in advance but allows himself some freedom of choice along the way in defining the individual terms of the sequence.

There seem to be at least two motivations for the introduction of free-choice sequences into the real number system. First, since each constructive real number can presumably be described by a phrase in the English language, on superficial consideration the set of constructive real numbers would appear to be countable. On closer consideration this is seen not to be the case: Cantor's uncountability theorem holds, in the following version. If $\{x_n\}$ is any sequence of real numbers, there exists a real number x with $x \neq x_n$ for all n. Nevertheless it appears that Brouwer was troubled by a certain aura of the discrete clinging to the constructive real number system $\mathbb{R}$. Second,

every function anyone has ever been able to construct from $\mathbb{R}$ to $\mathbb{R}$ has
turned out to be continuous, in fact uniformly continuous on bounded subsets.
(The function f that is 1 for $x \geq 0$ and 0 for $x < 0$ does not count, because
for those real numbers x for which we have no proof of the statement "$x \geq 0$,
or $x < 0$" we are unable to compute $f(x)$.) Brouwer had hopes of proving that
every function from $\mathbb{R}$ to $\mathbb{R}$ is continuous, using arguments involving free
choice sequences. He even presented such a proof [7]. It is fair to say that
almost nobody finds his proof intelligible. It can be made intelligible by replacing Brouwer's arguments at two critical points by axioms, that Kleene and
Vesley [21] call "Brouwer's principle" and "the bar theorem." My objection
to this is, that by introducing such a theorem as "all $f: \mathbb{R} \to \mathbb{R}$ are continuous" in the guise of axioms, we have lost contact with numerical meaning.
Paradoxically this terrible price buys little or nothing of real mathematical
value. The entire theory of free-choice sequences seems to me to be made of
very tenuous mathematical substance.

If it is fair to say that the intuitionists find the constructive concept of a
sequence generated by an algorithm too precise to adequately describe the real
number system, the recursive function theorists on the other hand find it too
vague. They would like to specify more precisely what is meant by an algorithm, and they have a candidate in the notion of a recursive function. They
admit only sequence of integers or rational numbers that are recursive (a concept we shall not define here: see [20] for details). Their reasons are, that
the concept is more precise than the naive concept of an algorithm, that every
naively defined algorithm has turned out to be recursive, and it seems unlikely
we shall ever discover an algorithm that is not recursive. This requirement
that every sequence of integers must be recursive is wrong on three fundamental grounds. First and most important, there is no doubt that the naive
concept is basic, and the recursive concept derives whatever importance it
has from some presumption that every algorithm will turn out to be recursive.
Second, the mathematics is complicated rather than simplified by the restriction to recursive sequences. If there is any doubt as to this, it can be resolved by comparing some of the recursivist developments of elementary analysis with the constructivist treatment of the same material. Even if one is
oriented to running material on a computer, the recursivist formulation would
constitute an obstacle, because very likely the recursive presentation would
be translated into computer language by first translating into common

constructive terminology (at least mentally) and then translating that into the language of whatever computer was being used. Third, no gain in precision is actually achieved. One of the procedures for defining the value of a recursive function is to search a sequence of integers one by one, and choose the first that is non-zero, having first proved that one of them is non-zero. Thus the notion of a recursive function is at least as imprecise as the notion of a correct proof. The latter notion is certainly no more precise than the naive notion of a (constructive) sequence of integers.

The desire to achieve complete precision, whatever that is, is doomed to frustration. What is really being sought is a way to guarantee that no disagreements will arise. Mathematics is such a complicated activity that disagreements are bound to arise. Moreover, mathematicians will always be tempted to try out new ideas that are so complicated or innovative that their meaning is questionable. What is important is not to develop some framework, such as recursive function theory, in the vain hope of forestalling questionable innovations, but rather to subject every development to intense scrutiny (in terms of the meaning, not on formal grounds).

Recursive functions come into their own as the source of certain counter-examples in constructive mathematics, the most famous being the word-problem in the theory of groups. Since the concept of a (constructively) recursive sequence is narrower than the concept of a (constructive) sequence, it is easier to demonstrate that there exist no recursive sequences satisfying a given condition G. Such a demonstration makes it extremely unlikely that a (constructive) sequence satisfying G will be found without some radically new method for defining sequences being discovered, a discovery that many view as almost out of the question.

Although some very beautiful counter-examples have been given by means of recursive functions, they have also been used as a source of counter-examples in many situations in which a prior technique due to Brouwer [20] would have been both simpler and more convincing. Brouwer's idea is to counter-example a theorem A by proving A → LPO. Since nobody seriously thinks LPO will ever be proved, such a counter-example affords a good indication that A will never be proved. As an instance, Brouwer has shown that the statement that every bounded monotone sequence of real numbers converges implies LPO.

Another source of Brouwerian counter-examples is the statement LLPO (for the "lesser limited principle of omniscience"), that if $\{n_k\}$ is any sequence of integers, then either the first non-zero term, if one exists, is even or else the first non-zero term, if one exists, is odd. Clearly LPO → LLPO, but there seems to be no way to prove that LLPO → LPO. Nevertheless, we are just as sceptical that LLPO will ever be proved. Thus A → LLPO is another type of Brouwerian counter-example for A. As an instance, the statement that "either $x \geq 0$ or $x \leq 0$ for each real number x" implies LLPO, in fact is equivalent to it.

Thus we are so sceptical that the statements LPO, LLPO, and their ilk will ever be proved that we use them for building counter-examples. The strongest counter-example to A would be to show that a proof of A is inconceivable, in other words to prove "not A," but proving "A → LPO" or "A → LLPO" is almost as good. In fact, I personally find it inconceivable that LPO (or LLPO for that matter) will ever be proved. Nevertheless I would be reluctant to accept "not LPO" as a theorem, because my belief in the impossibility of proving LPO is more of a gut reaction prompted by experience than something I could communicate by arguments I feel would be sure to convince any objective, well-informed, and intelligent person. The acceptance of "not LPO" as a theorem would have one amusing consequence, that the theorems of constructive mathematics would not necessarily be classically valid (on a formal level) any longer. It seems we are doomed to live with "LPO" and "there exists a function from $[0, 1]$ to $\mathbb{R}$ that is not uniformly continuous" and similar statements, of whose impossibilities we are not quite sure enough to assert their negations as theorems.

The classical paradoxes are equally viable constructively, the simplest perhaps being "this statement is false." The concept of the set of all sets seems to be paradoxical (i.e., lead to a contradiction) constructively as well as classically. Informed common sense seems to be the best way of avoiding these paradoxes of self reference. Their spectre will always be lurking over both classical and constructive mathematics. Hermann Weyl made the meticulous avoidance of self reference the basis of a whole new development of the real number system (see Weyl [32]) that has since become known as predicative mathematics. Weyl later abandoned his system in favor of intuitionism. I see no better course at present that to recognize that certain concepts are inherently inconsistent and to familiarize oneself with the dangers of self-reference.

Not only is there insufficient time, but I would not be competent to review all of the recent advances of constructive mathematics, including those of ad hoc constructivism as well as those taking place under constructivist philosophies at variance with those that I have presented here, for example the recursivist constructivism of Markov and his school in Russia. (I have been told that some of the recent advances in differential equations have tended to present that subject in a more constructive light. Perhaps Felix Browder will give us some information about those developments.) I shall restrict myself in what remains to selected developments with which I am familiar, that seem to me to be of special interest.

Brouwer [6] was the first to develop a constructive theory of measure and integration, and the intuitionist tradition (see [19] and [31] for instance) in Holland carried the development further, working with Lebesgue measure on $\mathbb{R}^n$. In [1] I worked with arbitrary measures (both positive and negative) on locally compact spaces, recovering much of the classical theory. The Daniell integral was developed in full generality in [5]. The concept of an integration space postulates a set X, a linear subset L of the set of all partially-defined functions from X to $\mathbb{R}$, and a linear functional I from L to $\mathbb{R}$ having the properties

(1) if $f \in L$, then $|f| \in L$ and $\min\{f, 1\} \in L$

(2) if $f \in L$ and $f_n \in L$ for each n, such that $f_n \geq 0$ and $\sum_{n=1}^{\infty} I(f_n)$ converges to a sum that is less than $I(f)$, then

$\sum_{n=1}^{\infty} f_n(x)$ converges and is less than $f(x)$, for some x in the common domain of f and the functions f_n

(3) $I(p) \neq 0$ for some $p \in L$

(4) $\lim_{n \to \infty} I(\min\{f, n\}) = I(f)$ and $\lim_{n \to \infty} I(\min\{|f|, n^{-1}\}) = 0$ for all f in L.

We define L_1 to consist of all partially defined functions f from X to $\mathbb{R}$ such that there exists a sequence $\{f_n\}$ of elements of L such that

(a) $\sum_{n=1}^{\infty} I(|f_n|)$ converges and (b) $\sum_{n=1}^{\infty} f_n(x) = f(x)$ whenever $\sum_{n=1}^{\infty} |f_n(x)|$ converges.

It turns out to be possible to extend I to L_1, in such a way that (X, L_1, I) also satisfy the axioms, and in addition L_1 is complete under the metric $\rho(f, g) = I(|f - g|)$.

The only real problem in recovering the classical Daniell theory is posed by the classical result that if $f \in L_1$ then the set $A_t = \{x \in X : f(x) \geq t\}$ is integrable for all $t > 0$ (in the sense that its characteristic function χ_t, defined by $\chi_t(x) = 1$ if $f(x) \geq t$ and $\chi_t(x) = 0$ if $f(x) < t$, is in L_1). The constructive version is that A_t is integrable for all except countably many $t > 0$. The proof of this requires a rather complex theory, called the theory of profiles. Y. K. Chan informs me that he has been able to simplify the theory of profiles considerably. He has also effected a considerable simplification in another trouble-spot of [5], the proof that a non-negative linear functional I on the set $L = C(X)$ of continuous functions on a compact space X satisfies the axioms for an integration space presented above. (Axiom (2) is the troublemaker.)

Constructive integration theory affords the point of departure for some recent constructivizations of parts of probability theory. There is no (constructive) way to prove even the simplest cases of the ergodic theorem, such that if T denotes rotation of a circle X through an angle α, then for each integrable function $f : X \to R$ and almost all x in X, the averages

$$f_N(x) = \frac{1}{N} \sum_{n=1}^{N} f(T^n x)$$

converge. (The difficulty comes about because we are unable to decide for instance whether $\alpha = 0$.) One way to recover the essence of the ergodic theorem constructively, and in fact deepen it considerably, is to show that the sequence $\{f_N\}$ satisfies certain integral inequalities, analogous to the up-crossing inequalities (see [14]) of martingale theory. This was done in the context of the Chacon-Ornstein ergodic theorem in [1], and even more generally in [3].

John Nuber [23] takes another route. He presents sufficient conditions, close to being necessary, that the sequence $\{f_N\}$ actually converges a.e., in the context of the classical Birkhoff ergodic theorem. More recently, in an unpublished manuscript, he has generalized his conditions to the context of the classical Chacon-Ornstein theorem.

Y. K. Chan has done much to constructivize the theory of stochastic processes. His paper [10] unifies the two classically distinct cases of the renewal theorem into one constructive result. His paper [12] contains the following theorem:

THEOREM. Let μ_1 and μ_2 be probability measures on R, and f_1 and f_2 their characteristic functions (Fourier transforms). Let g be a continuous function on R, with $|g| \leq 1$. Then for every $\epsilon > 0$ there exist $\delta > 0$ and $\theta > 0$, depending only on ϵ and the moduli of continuity of f_1, f_2, and g, such that

$$\left| \int g d\mu_1 - \int g d\mu_2 \right| < \epsilon$$

whenever $|f_1 - f_2| < \delta$ on $[-\theta, \theta]$.

A simple corollary is Levy's theorem, that if $\{\mu_n\}$ is a sequence of probability measures on R, whose characteristic functions $\{f_n\}$ converge uniformly on compact sets to some function f, then μ_n converges weakly to a probability measure μ whose characteristic function is f

Levy's theorem is classically an important tool for proving convergence of measures. Chan shows that this is also true constructively, by using it to get constructive proofs of the central limit theorem and of the Levy-Khintchine formula for infinitely divisible distributions.

Chan's papers [9] and [11] are primarily concerned with the problem of constructing a stochastic process. In [9] he gives a constructive version of Kolmogorov's extension theorem. In [11], he constructivizes the classical derivation of a time homogeneous Markov process from a strongly continuous semi-group of transition operators. In addition he proves Ionescu Tulcea's theorem and a supermartingale convergence theorem.

H. Cheng [13] has given a very pretty version of the Riemann mapping theorem and Caratheodory's results on the convergence of mapping functions. He defines a simply connected proper open subset U of the complex plane $\mathbb{C}$ to be __mappable__ relative to some distinguished point z_0 of U if for each $\epsilon > 0$ there exist finitely many points $z_1, \ldots, z_n$ in the complement of U such that any continuous path beginning at z_0 and having distance $\geq \epsilon$ from each of the points $z_1, \ldots, z_n$ lies entirely in U. He shows that mappability is necessary and sufficient for the existence of a mapping function. He goes on to study the dependence of the mapping function on the domain, by defining

natural metrics on the space D of domains (with distinguished points z_0) and
the space M of mapping functions, and proving that the function $\lambda : D \to M$
that associates to each domain its mapping function is a homeomorphism. He
thus extends and constructivizes the classical Caratheodory results. Many of
his estimates are similar to those developed by Warschawski in his studies of
the mapping function.

 The problem of constructivizing the classical theory of uniformization
is still open. (Even reasonable conjectures seem difficult to come by.) So is
the problem of (constructively) constructing canonical maps for multiply connected domains, as far as I know.

 It is natural to define two sets to have the same cardinality if they are
in one-one correspondence. The constructive theory of cardinality seems
hopelessly involved, due to the constructive failure of the Cantor-Bernstein
lemma, and for other reasons as well.

 Progress has been made however in constructivizing the theory of
ordinal numbers. Brouwer [8] defines ordinals to be ordered sets that are
built up from non-empty finite sets by finite and countable addition.
F. Richman [26] gives a more general definition. Simple in appearance, his
definition constructivizes the property of induction in just the right way. An
ordinal number (or well-ordered set) is a set S with a binary relation < such
that

(1)   if $a < b$ and $b > c$, then $a < c$

(2)   one and only one of the relations $a < b$, $b < a$, $a = b$ holds for given
      elements a and b of S

(3)   let T be any subset of S with the property that every element b of
      S, such that $a \in T$ for each a in S with $a < b$, belongs to T; then
      T = S.

 Richman shows that each Brouwerian ordinal satisfies (1), (2), and (3).
He gives examples of ordinals (in his sense) that are not Brouwerian. He
shows that every subset of an ordinal is an ordinal (under the induced order).
He uses his theory to constructivize the classical theorems of Zippin and Ulm
concerning existence and uniqueness of p-groups with prescribed invariants.

 The above examples might give the impression that the constructivization of classical mathematics always proceeds smoothly. I shall now give
some other examples, to show that in fact it does not.

In [1] the Gelfand theory of commutative Banach algebras was con-
structivized to a certain extent. The theory has to be considered unsatis-
factory, not because the classical content is not recovered (it is), but because
it is so ugly. It is almost certain that a prettier constructivization will some-
day be found.

Stolzenberg [28] gives a meticulous analysis of some of the considerations involved in constructivizing a particular classical theory, the open mapping theorem and related material. Again, an incisive constructivization is
not obtained.

J. Tennenbaum [29] gives a deep and intricate constructive version of
Hilbert's basis theorem. Consider a commutative ring A with unit. It would
be tempting to call A (constructively) Noetherian if for each sequence $\{a_n\}$
of elements of A there exists an integer N such that for $n \geq N$ the element
a_n is a linear combination of $a_1, \ldots, a_{n-1}$ with coefficients in A. This
notion would be worthless -- not even the ring of integers is Noetherian in this
sense. In case A is discrete (meaning that the equality relation for A is
decidable), the appropriate constructive version of Noetherian seems to be the
following (as given in [29]).

DEFINITION. A sequence $\{a_n\}$ of elements of A is almost eventually zero if for each sequence $\{n_k\}$ of positive integers there exists a positive integer k such that $a_n = 0$ for $k \leq n \leq k + n_k$.

DEFINITION. A basis operation r for A is a rule that to each finite
sequence $a_1, \ldots, a_n$ of elements of A assigns an element $r(a_1, \ldots, a_n)$ of
A of the form $a_n + \lambda_1 a_1 + \cdots + \lambda_{n-1} a_{n-1}$, where each λ_i belongs to A.

DEFINITION. A is Noetherian if it has a basis operation r such that
for each sequence $\{a_n\}$ of elements of A the associated sequence
$\{r(a_1, \ldots, a_n)\}$ is almost eventually zero.

Tennenbaum proved the appropriateness of his version of Noetherian by
checking out the standard cases and proving the Hilbert basis theorem. He
also extended his definition and results to the case of a not-necessarily discrete ring A. The theory in that case is so complex that it cannot be considered satisfactory.

In spite of the pioneering efforts of Kronecker, and continued work by
many algebraists, resulting in many deep theorems, the systematic

constructivization of algebra would seem hardly to have begun. The problems
are formidable. A very tentative suggestion is that we should restrict our
attentions to algebraic structures endowed with some sort of topology, with
respect to which all operations and maps are continuous. The work of
Tennenbaum quoted above might provide some ideas of how to accomplish this.
The task is complicated by the circumstance that no completely suitable constructive framework for general topology has yet been found.

The constructivization of general topology is impeded by two obstacles.
First, the classical notion of a topological space is not constructively viable.
Second, even for metric spaces the classical notion of a continuous function is
not constructively viable; the reason is that there is no constructive proof that
a (pointwise) continuous function from a compact (complete and totally bounded)
metric space to $\mathbb{R}$ is uniformly continuous. Since uniform continuity for
functions on a compact space is the useful concept, pointwise continuity (no
longer useful for proving uniform continuity) is left with no useful function to
perform. Since uniform continuity cannot be formulated in the context of a
general topological space, the latter concept also is left with no useful function to perform.

In [1] I was able to get along by working mostly with metric spaces and
using various ad hoc definitions of continuity: one for compact spaces, another for locally compact spaces, and another for the duals of Banach spaces.
The unpublished manuscript [4] was an attempt to develop constructive general
topology systematically. The basic idea is that a topological space should consist of a set X, endowed with both a family of metrics and a family of boundedness notions, where a boundedness notion on X is a family S of subsets
of X (called bounded subsets), whose union is X, closed under finite unions
and the formation of subsets.

For example, let C be the set of all real valued functions f: $\mathbb{R} \to \mathbb{R}$,
bounded and (uniformly) continuous on finite intervals. Each finite interval
of $\mathbb{R}$ induces a metric on C (the uniform metric on that interval). In addition, there is a natural boundedness notion S. A subset E of C belongs to
S if there exists $r > 0$ such athat $|f| \le r$ for all f in E. A sequence $\{f_n\}$
of elements of C <u>converges</u> to an element f of C if it converges with respect to each of the metrics on C, and if it is bounded.

The notion of a continuous function from one such space to another, as
given in [4], is somewhat involved and will not be repeated here. It was

possible to develop a theory that seems to accommodate the known examples
and to have certain pleasing functorial qualities, but the theory is somehow not
convincing -- for one thing, it is too involved. For another, there is a certain
sort of space -- let us call it a ball space -- that does not fit well into the
theory.

DEFINITION. A ball space is a set X, together with a function that to
each $r \geq 0$ and point x of X associate a subset $B(x, r)$ of X (to be thought
of as the closed ball of radius r about x) satisfying the following axioms.

(a) $B(x, r) \subset B(x, s)$ if $r \leq s$.

(b) $B(x, 0) = \{x\}$.

(c) $B(x, r) = \cap \{B(x, s) : s > r\}$.

(d)   If $y \in B(x, r)$, then $x \in B(y, r)$.

(e)   If $y \in B(x, r)$ and $z \in B(y, s)$, then $z \in B(x, r + s)$.

(f) $\cup \{B(x, r) : r \geq 0\} = X$.

Duals of Banach spaces are particular instances of ball spaces, as are
various other function spaces.

Algebraic topology, at least at the elementary level, should not be too
difficult to constructivize. There is a problem with defining singular cohomology constructively, as pointed out in [2]. Richman [25] points out that
the classical Vietoris homology theory is not satisfactory constructively, and
he gives a new version that constructively (and also classically) has certain
features that are more desirable.

I would like to conclude these lectures by discussing some of the tasks
that face constructive mathematics.

Of primary importance is the systematic constructive development of
enough of algebra for a pattern to begin to emerge. Of course, it may be that
much of the classical theory is inherently unconstructivizable, and that constructive algebra will go its own way. It is too early to tell.

Less critical, but also of interest, is the problem of a convincing constructive foundation for general topology, to replace the ad hoc definitions in
current use. It would also be good to see a constructivization of algebraic
topology actually carried through, although I suspect this would not pose the
critical difficulties that seem to be arising in algebra.

To sum up, the first task is to constructivize as much of existing
classical mathematics as is suitable for constructivization. As this is being

done, we should increasingly turn our attention to questions of the efficiency of
our algorithms, and bridge the gap between constructive mathematics on the
one hand and numerical analysis and the theory of computation on the other.
Since constructive mathematics is the study of what is theoretically computable, it should afford a sound philosophical basis for the theory of computation.

Our terminology and technical devices need constant re-examination as
to whether they are the most appropriate tools for extracting the full meaning
from our material. It seems to me that the meaning of implication, in particular, should be thoroughly studied, and other possible candidates investi-
gated. Such statements as "$(A \to B) \to C$" have a rather tenuous meaning, and
in many instances of proofs of such statements, something more is actually
being proved. Work of Gödel [17] raises some interesting possibilities about
possible re-definitions of implication, which seem to be very difficult to imple-
ment in usable generality, and which also seem to run counter to natural
modes of thought. There seems to be no reason in principle that we should
not be able to develop a viable terminology that incorporates more than one
meaning for some or all of the quantifiers and connectives.

More important than any of these technical problems is the broader
problem of involving ourselves more deeply with the meaning of mathematics
at all levels. This is the simplest and most general statement of the constructivist program, and the technical developments are intended as a means
to that end.

REFERENCES

1. Bishop, E., "Foundations of Constructive Analysis," McGraw-Hill, New York, 1967.

2. Bishop, E., Mathematics as a Numerical Language, in "Intuitionism and Proof Theory," North-Holland, Amsterdam, 1970.

3. Bishop, E., A Constructive Ergodic Theorem, J. Math. Mech., 17 (1968), 631-640.

4. Bishop, E., Notes on Constructive Topology, UCSD, 1971.

5. Bishop, E. and H. Cheng, "Constructive Measure Theory," AMS Memoir 116, Providence, 1972.

6. Brouwer, L. E. J., Begründung der Mengenlehre Unabhängig vom Logischen Satz vom Ausgeschlossenen Dritte II: Theorie der Punktmengen, Ver. Kon. Akad. Wet. (Amsterdam), 12 (1919), 1-33.

7. Brouwer, L. E. J., Beweiss dass Jede Volle Funktion Gleichmässig Stetig ist, Kon. Ned. Akad. Wet. (Amsterdam), 27 (1924), 189-193.

8. Brouwer, L. E. J., Zur Begründung der Intuitionistischen Mathematik III, Math. Ann., 96 (1927), 451-488.

9. Chen, Y. K., A Constructive Approach to the Theory of Stochastic Processes," Trans. AMS, 165 (1972), 37-44.

10. Chan, Y. K., A Constructive Version of the Renewal Theorem, to appear.

11. Chan, Y. K., Notes on Constructive Probability Theory, to appear.

12. Chan, Y. K., On Constructive Convergence of Measures on the Real Line, to appear.

13. Cheng, H., Ph.D. Dissertation, University of California, San Diego, 1971.

14. Doob, J. L., "Stochastic Processes," Wiley, New York, 1953.

15. Frankel, A. and Y. Bar-Hillel, "Foundations of Set Theory," North-Holland, Amsterdam, 1958.

16. Gödel, K., What is Cantor's Continuum Problem?" Amer. Math. Monthly, 54 (1947), 515-525.

17. Gödel, K., Uber Eine Bisher Noch Benutzte Erweiterung des Finiten Standpunktes, Dialectica, 12 (1958), 280-287.

18. Goodman, N. D. and J. Myhill, The Formalization of Bishop's Constructive Mathematics, in "Topics in Algebraic Geometry, and Logic," Springer, Berlin, 1972.

19. Heyting, A., Note on the Riesz-Fischer Theorem, Proc. Kon. Akad. Wet. (Amsterdam), 54 (1951), 35-40.

20. Kleene, S. C., "Introduction to Metamathematics," Van Nostrand, Princeton, 1952.

21. Kleene, S. C. and R. Vesley, "The Foundations of Intuitionistic Mathematics," North-Holland, Amsterdam, 1965.

22. Myhill, J., Formal Systems of Intuitionistic Analysis III-Lawlike Analysis, to appear.

23. Nuber, J., A Constructive Ergodic Theorem, Trans. AMS, 164 (1972), 115-137.

24. Nuber, J., A Constructive Proof of the Chacon-Ornstein Theorem, to appear.

25. Richman, F., A Constructive Modification of Vietoris Homology, to appear.

26. Richman, F., The Constructive Theory of Countable Abelian P-Groups, to appear in Pacific Jour. Math.

27. Spector, C., Provably Recursive Functionals of Analysis, in Amer. Math. Soc. Symposium on Pure Mathematics, No. 5, Providence, 1962, 1-27.

28. Stolzenberg, G., A Critical Analysis of Banach's Open Mapping Theorem, to appear.

29. Tennenbaum, J., Ph.D. Dissertation, University of California, San Diego, 1973.

30 van der Meulen, S. G. and C. H. Lindsey, "Informal Introduction to Algol 68," Math. Centrum, Amsterdam, 1969.

31. van Rootselaar, B., Generalization of the Brouwer Integral, Ph.D. Dissertation, Amsterdam, 1954.

32. Weyl, H., "Das Kontinuum," Leipzig, 1918. Reprinted by Chelsea, New York, 1932.

Contemporary Mathematics
Volume 39, 1985

ERRETT BISHOP - IN MEMORIAM

Stefan E. Warschawski

It was a sad day last April 14, when we heard the - not entirely unex-
pected - news that Errett Bishop was no longer among us. My thoughts went
back 18 years ago, to the time he came to U.C.S.D. I vividly recalled how
jubilant we all were when we learned that Errett had accepted our invitation to
join our fledgling department! Sometime early in the year 1965 we heard the
rumor that Errett Bishop was leaving Berkeley and that a number of universities, Yale, Brown, the University of Chicago, Wisconsin and others, were
making efforts to get him. When we attempted to call him in Berkeley he was
not there, and we finally reached him in New Haven where he was visiting the
Yale mathematics department. He promised to honor our request that he
make no commitment to Yale or other universities before he had an opportunity to visit us. He kept his promise, and after the official appointment
formalities were completed, accepted the offer from U.C.S.D. We all in the
department realized the stroke of good luck we had in attracting a colleague
of his ability and stature - for our personal benefit and the prestige his coming
gave to our young department. For already then Errett Bishop was a mathematician of international renown.

Errett Bishop was born in Newton, Kansas, July 24, 1928. He showed
his ability early. He was not challenged by his high school work, and when he
learned of a special scholarship program at the University of Chicago, he
applied and was accepted at the age of 16. He received the B.S. at the age of
19 and his M.S. degree two years later. He was a brilliant student. Paul
Halmos, who later was his advisor for the Ph.D., recalls that Errett took
Halmos' graduate course in probability as a sophomore. As an exercise
Halmos suggested once to the class that they find one example of a certain
unusual phenomenon. Bishop was not satisfied to do this; his paper contained
a general theorem, a necessary and sufficient condition for that phenomenon
to occur. The theorem was new to Halmos.

From 1950-52 Errett served in the U.S. Army. The Army showed good judgment by assigning him (after basic training) to do mathematical research at the National Bureau of Standards. This work was recognized by an award: he received a "Commendation ribbon with pendant for mathematical research in Army Ordnance."

In 1952 he returned to the University of Chicago and obtained his Ph.D. in 1954. Halmos, his advisor, said: "As a student he was outstanding and with the writing of his thesis he became spectacular." His thesis was a penetrating study of a generalized version of spectral theory.

Errett began his teaching career at Berkeley in 1954, where he remained until 1965, with the exception of 1 1/2 years spent at the Institute for Advanced Study in Princeton as a member. While at Berkeley he had a Sloane Fellowship for three years, a high distinction given to young scholars of unusual ability, and he spent the years 1964-65 as a member of the Miller Institute for Basic Research in Berkeley - allowing him to devote full time to research (in preparation for his remarkable book, which I describe later on). In his first 8 years at Berkeley he advanced from instructor to a full professor in their distinguished mathematics department.

Errett Bishop is universally recognized as an outstanding mathematician, one of the great analysts of the present time. All colleagues, who know his work, marvel at the power and concentration he was able to bring to bear on the problems on which he worked and the originality, depth and insight he displayed in his work. I like to mention some of the areas of mathematics to which Errett has made essential contributions. This will be meaningful only to the mathematicians in the audience; to others it might give an idea of the scope and diversity of his research. We have five speakers here today who are experts in the various areas to which Errett contributed and will speak in more detail about his work.

Many mathematicians earn their reputation by doing outstanding work in one area, but Errett Bishop made his mark in many. He had the ability of going into a new field and in two or three papers advancing it in an essential way. He has many times succeeded in unifying and extending a field by introducing the right kind of imaginative new concepts. He did that first in operator theory in Hilbert space and in Banach space; next in the theory of polynomial approximation in the complex plane and on Riemann surfaces. The latter led Errett to his outstanding work in function algebras. This work in turn led him to his highly original approach to the theory of functions of several complex variables. He applied here the methods of functional analysis as a new and powerful tool. This work had a strong impact on the theories of mathematicians in Europe.

There are theorems and methods in all of these areas which bear his name and are used by many researchers. He had a seminal influence in several directions.

About 1964 Errett turned his interests toward the foundations of mathematics. Despite his extraordinary successes in so many different fields - which brought him recognition and acclaim of his peers - his independent mind drove him to think about the basic meaning of it all. He felt that there is a crisis in mathematics, due to our neglect of philosophical issues. His motivation was to explore the meaning of mathematics: he felt that we are proving all those theorems - more today than ever before; mathematics, in that sense, flourishes as never before - without really knowing what they mean. He asked: "Is pure mathematics simply a game which we play or do our theorems describe an external reality?" A basic point is already made by asking "what is an integer?" To quote an example he gave in one of his general lectures: The integer 3 has a different quality from the integer

$$3^{99^{99^{99}}}$$

which in turn differs in quality from the integer defined in the following way: $N = 0$ if an as yet unsettled mathematical conjecture is false and $N = 1$ if it is true (for example, the Goldbach conjecture that every even integer $\geqq 4$ is the sum of two primes).

In the first case the integer 3 may be clearly computed; in the second case we can compute the integer in principle. In the third case we cannot find the answer. Classically all three integers are "defined."

The constructive point of view is that an integer be defined only if it can be calculated in a finite number of steps. According to Bishop the basic constructivist goal is that all mathematics concern itself with the precise description of finitely performable (abstract) procedures.

In his ideas of constructivism Errett Bishop goes back to the Dutch mathematician J. L. E. Brouwer (1881-1967) and to L. Kronecker (1828-1891). Brouwer's theory, called Intuitionism, stresses as basic our intuition of the integers and of the real numbers. All of mathematics is to be reduced to these two primitive constructs. Bishop proposes (following Kronecker rather than Brouwer) that the integers are the only basic (irreducible) mathematical construct. In his book "Foundations of Constructive Analysis" which appeared early in 1968, he not only introduced the fundamental concepts of his constructive mathematics, but he actually developed a large part of contemporary (modern) analysis by constructive methods - or, more precisely, a constructive version of a large part of modern analysis. You only need to look at the table of contents to perceive the enormity of this work. The book not only shows his superb mastery of so many fields; by studying it you realize the tremendous achievement in rethinking these fields under the rules of constructivism. In reading it the "classical" expert in a particular area discovers original, novel methods which can also be used to improve classical (conventional) proofs of many theorems.

Whenever philosophical issues are introduced, there often arise disagreements. Brouwer's view had support from some of the great mathematicians of our time, such as H. Weyl and H. Poincaré. In the early 1930's-1931 or 1932, Professor Weyl taught the course in Calculus at the University of Göttingen from the intuitionist point of view. I remember as a young postdoctoral in Gottingen at the time, sitting in some of the lectures. D. Hilbert disagreed with Brouwer.

When Errett's book came out, there were enthusiastic supporters of Errett's point of view. They felt that his book was the first really important work on this subject since that of Brouwer. Actually, it went far beyond Brouwer. It demonstrates clearly that the constructivisation of mathematics can be done, in doing so, our understanding of mathematics is deepened. One analyst wrote: "It is no exaggeration to call this book one of the truly significant mathematical works of this century."

Many mathematicians consider themselves as "practicing" mathematicians and are not much involved with the foundations. But they feel that when a man of such depth and imagination turns to constructive mathematics, then mathematicians should no longer disregard this field.

Some mathematicians do not agree with Bishop that there is a crisis in mathematics, and some who may, do not feel it necessary to adopt the constructivist point of view as a way out of that crisis. Although Bishop has shown that much of contemporary mathematics can be done in this way - and continued to "constructivise" further fields - they may feel that too much of good mathematics, accomplished by conventional methods, may have to be abandoned by insisting on constructively valid proofs. But even these mathematicians admire Errett's book for the enormous intellectual accomplishment it represents. The late Professor A. Robinson of Yale (and earlier U.C.L.A.) who introduced "Non-standard Analysis," a theory which has been described as diametrically opposed to constructivism, wrote "Even those who are not willing to accept Bishop's basic philosophy must be impressed with the great analytical power displayed in his work."

There is another, less basic and more pragmatical, aspect to the constructivist approach. In certain areas, such as differential equations, integral equations, complex analysis, and others, it has become the goal of some mathematicians to obtain "constructable" (computable) solutions (within the framework of "classical" mathematics) as opposed to abstract demonstration of their existence. This had been particularly stimulated by the development of high speed computers. There are monographs on "Constructive Function Theory," "Constructive Methods in Conformal Mapping" and others in the western mathematical literature and in Russia. These deal mainly with approximation theories and are also of interest in numerical analysis and in applied mathematics. Errett's book - although not intended for this purpose - will undoubtedly have a stimulating influence also in that direction.

Errett Bishop received much recognition of his work through invitations to lecture about it at national and international meetings, at symposia in the various areas of his interests, and at mathematics colloquia of numerous universities. Since 1959 there was a steady stream of such invitations. I shall not mention them all - I do not know all of them. Let me mention some of the most prestigious: In 1966 he was asked to give an address at the International Congress of Mathematicians which met in Moscow that year. (The Congress is held every 4 years, each time in a different country.) He was one of only 22 mathematicians from the U.S. selected for an invited address by the international committee; it was estimated that over 500 mathematicians from the U.S. presented papers, and many more attended the Congress. The title of his talk was: "The Constructivisation of Abstract Analysis" and was the first opportunity for him to present his ideas to an international audience.

Both American mathematical organizations, the Mathematical Association of America and the American Mathematical Society, honored him with their highest honors in lectureships. (For the uninitiated, the former is primarily concerned with college teaching and mathematical research of more immediate benefit to educators, while the latter is mainly concerned with research.) The Association invited Errett Bishop to deliver the distinguished Hedrick Lectures at their 1969 Summer Meeting, at the University of Oregon.

The American Mathematical Society invited Errett to give the prestigious Colloquium Lectures at its summer meetings in 1973. Such invitations are extended only to individuals who have developed a substantial body of new mathematics. It is a four-lecture series, each of one hour length. (The title of Errett's lectures was "Schizophrenia of Contemporary Mathematics.") This invitation recognized the importance of presenting Errett Bishop's theory of constructive mathematics to a larger mathematical public concerned with research.

In addition to these special lecture series both organizations sponsor one-hour addresses. Errett received numerous invitations to give these, at national and regional meetings.

I like to mention two symposia in which he participated. One was a Symposium on Functions of Several Complex Variables, at the University of Minnesota in 1964 where he spoke about his famous work, and the other, The 10th Holiday Mathematics Symposium, December 27-31, 1972, at the New Mexico State University, centered around him as the principal lecturer with an opportunity for the participants for discussion and interchange of ideas. The topic was "Aspects of Constructivism."

I believe that Errett's popularity as a lecturer was not only due to what he had to say, but also to how he said it. He was a marvelous lecturer. The clear insight which he possessed enabled him to speak simply, avoid sidetracks, make it easy to follow him. It was a pleasure to listen to him. This ability made him a superb teacher. When he first came to U.C.S.D. he taught only on the graduate level. He had a number of research students at U.C.S.D. (8 of whom received the Ph.D. under his direction). Whatever course he taught - he had always a highly original, individualistic approach; he devised interesting and challenging problems, always stimulating the imagination of his students - and speaking of my own experience - his colleagues as well.

He was always interested in the undergraduate curriculum; he was one of the first senior members in the department who taught freshman calculus.

He was a compassionate teacher, had a strong feeling for his students; he did not want them to feel undue pressure and made himself available to help them in their work. The students liked him very much; they appreciated his efforts and his sense of fairness.

In 1969 Errett was elected to the American Academy of Arts and Sciences.

Errett was a remarkable personality. Particularly outstanding traits were his independence and originality, apparent in everything he did, in his research, his teaching, in every aspect of daily life He had strong principles by which he lived and a strong feeling for fairness in the treatment of other people - his colleagues, his students. He treated people with kindness and consideration.

He was a very private person and did not talk much about himself. For example, it was only recently that I learned that he had a stamp collection. I saw in the house of Jane and Errett Bishop a beautiful collection of Indian and Mexican artifacts, but did not realize that he was the ardent collector. He had many other cultural interests.

Errett's untimely death brought a grievous loss. Mathematical science and our department have lost a great man, and many of us in the department a sincere and compassionate friend.

Contemporary Mathematics
Volume 39, 1985

THE WORK OF ERRETT BISHOP ON SEVERAL
COMPLEX VARIABLES

John Wermer

Errett Bishop's work on the function theory of several complex variables
is fundamental. In a few years in the early 1960's he obtained a series of results which have given new directions to the subject, and which form the basis
for important research being done at the present time.

Bishop's methods were quite personal to him. They were direct, elementary and powerful.

In the present article we shall try to give a brief indication of five major
problems which he tackled, and state a few of his main results. We shall also
give references to some of the continuing work on these questions done by
other mathematicians.

1. ANALYTIC POLYHEDRA

Let D be a simply connected domain in $\mathbb{C}$ and K a compact subset of
D. In his "lemniscate theorem" of 1897 Hilbert showed that we can insert in
D a lemniscate: $|P(x)| = R$, with P a polynomial in z and R a constant,
which surrounds K. In other words, a suitable component K_p of the set
$\{z \in \mathbb{C} \mid |P(x)| \leq R\}$ satisfies: $K \subset K_p \subset D$.

In [1] Bishop proved a theorem of this type in n dimensions. Let D
be a domain in $\mathbb{C}^n$. By an <u>analytic polyhedron</u> in D, globally defined, we
mean a set π which is the union of connected components of the set

$$\{z \in D \mid |f_i(z)| < 1, \quad i = 1, 2, \ldots, k\}$$

where $f_1, \ldots, f_k$ are holomorphic functions on D. If $k = n$, π is called a
<u>special analytic polyhedron</u>.

THEOREM 1. [1] Let D be a domain in $\mathbb{C}^n$ with D holomorphically convex. Let K be a compact subset of D. Then $\exists$ a special analytic polyhedron π in D such that $K \subset \pi \subset D$.

The analogous statement is true when D is replaced by an n-dimensional Stein manifold. This fact is a key ingredient in Bishop's proof of the embedding theorem for Stein manifolds. Hassler Whitney had shown that each s-dimensional (real) smooth manifold admits a smooth embedding as a smooth submanifold of R^{2s+1}. Using Theorem 1, Bishop proved the following: let X be a Stein manifold of complex dimension n. Then $\exists$ a closed analytic submanifold $\tilde{X}$ of $\mathbb{C}^{2n+1}$ such that X and $\tilde{X}$ are equivalent under a biholomorphic map. This includes the remarkable fact that each plane domain, each open Riemann surface, and each holomorphically convex domain in $\mathbb{C}^n$ can be realized as a closed analytic submanifold of some higher dimensional complex Euclidean space.

This embedding theorem for Stein manifolds was originally stated by Remmert, and proved by R. Narasimhan in [13] and by Bishop in [1]. It has been the source of much further work. It allowed Andreotti and Frankel in [3] to give a proof of the vanishing of the cohomology groups $H^P(X, \mathbb{C})$ for $p > n$ for each Stein manifold X, by embedding X in some $\mathbb{C}^N$ and using Morse theory.

Bishop himself was able to use Theorem 1 in [2] and [3] to give a new proof of the basic extension theorem for a function defined and analytic on an analytic subset of a Stein manifold, and of related results.

2. EXISTENCE OF ANALYTIC DISKS

Certain sets K in $\mathbb{C}^n$, $n > 1$, have an automatic analytic extension property, in the sense that there exists an open set U such that every function analytic in some neighborhood of K, no matter how small, extends analytically to U.

In particular, if Ω is a bounded domain in $\mathbb{C}^n$, $n > 1$, with connected boundary bdΩ, then bdΩ has this property, with $U = \Omega$. This fact was discovered by Hartogs in the first decade of this century.

bdΩ has real codimension 1 in $\mathbb{C}^n$. What if M is a smooth k-dimensional submanifold of $\mathbb{C}^n$ with codim(M) > 1? H. Lewy, in [12], was the first to exhibit such an extension phenomenon. He gave an example with k = 4, n = 3.

In [4] Bishop made a decisive contribution to this question. Let M be a smooth submanifold of $\mathbb{C}^n$ of real dimension k. Assume that $k \geq n$. Bishop gave a method for the construction of <u>analytic disks</u>: $z = z(\varsigma)$, $|\varsigma| \leq 1$, where $\varsigma \to z(\varsigma)$ is analytic, such that the boundary of the analytic disk: $\{z(\varsigma) \mid |\varsigma| = 1\}$ lies on M. Because of the Kontinuitäts-Satz, the existence of sufficiently many such disks can be used to prove an automatic analytic extension property for M. Furthermore, every single such disk belongs to the polynomially convex hull of M.

In the simplest case of Bishop's construction $k > n$ and we fix a point P of M at which a certain non-degeneracy condition holds. In a neighborhood of P, then M can be parametrically represented as follows:

(1)
$$z = (x + ih(x, w), w)$$

where $x \in R^{2n-k}$, $w \in \mathbb{C}^{k-n}$, and h is a smooth function of x and w taking values in R^{2n-k} and vanishing to the second order at $(0, 0)$. The local parameter space for M thus is $R^{2n-k} \times \mathbb{C}^{k-n} \simeq R^k$.

For each L^2-function $x : \varsigma \to x(\varsigma)$ defined on $|\varsigma| = 1$ and taking values in R^{2n-k}, we denote by Tx the conjugate function of x so that $x + iTx$ is the boundary value of some analytic function w on $|\varsigma| < 1$ whose imaginary part vanishes at $\varsigma = 0$. We fix arbitrarily a function w on $|\varsigma| = 1$ with values in $\mathbb{C}^{k-n}$ which is the boundary value of a function analytic in $|\varsigma| < 1$. Suppose now that we can find a function x such that

(2)
$$Tx = h(x, w)$$

as a function on $|\varsigma| = 1$. We put

$$z = (x + iTx, w) \text{ on } |\varsigma| \le 1 .$$

On $|\varsigma| = 1$, in view of (2), then,

$$z = (x + ih(x, w), w) .$$

By the parametric representation (1), z maps $|\varsigma| = 1$ to a closed curve on M. Also, since $x + iTx$ and w are analytic on $|\varsigma| < 1$, z maps $|\varsigma| \le 1$ onto an analytic disk in $\mathbb{C}^n$. This analytic disk then has its boundary on M.

To solve the equation (2) for x, a non-linear problem, Bishop used a fixed-point argument in function-space.

The case $k = n$ is harder. A modified form of the construction was carried out by Bishop for certain "exceptional" points P on M, at which the tangent space to M contains a complex line. Such points are, in general, isolated points on M, or absent altogether. This raises the global question:

given a smooth compact k-manifold in C^n, when do $\exists$ exceptional points of
the required type on M? In a special case, Bishop was able to use a topological result of Chern and Spanier to prove:

THEOREM: In general, a smooth 2-sphere M^2 in C^2 has on it at least two
suitable exceptional points, near which there exist families of analytic disks
with boundary curves lying on M^2.

This result gives a geometric "explanation," for a class of cases, of
Browder's theorem in [7] that a compact orientable 2-manifold in C^2 is
never polynomially convex. The higher-dimensional analogues of this result,
as well as the global problem of determining all analytic disks whose boundaries lie on a given M, are difficult and have led to much research ([6],
[20], etc.)

3. ALGEBRAS OF HOLOMORPHIC FUNCTIONS

In many problems of complex analysis in several variables one is required to know the spectrum of certain algebras of holomorphic functions. In
[5] and [6] Bishop studied the structure of such spectra.

Consider an algebra β of holomorphic functions on a connected com-
plex manifold X. For each compact set K in X, put $\|f\|_K =$
$\max\{|f(p)| \mid p \in K\}$. This makes β into a normed algebra. By the
spectrum S of β relative to this norm we mean the set of homomorphisms
ϕ of $\beta \to C$ with $|\phi(x)| \leq \|x\|_K$ for all x in β.

Assume now that X has complex dimension n and that for each point
p in X, β contains an n-tuple of functions giving local coordinates at p.

THEOREM 2: [5] Let β, $\| \|_K$, S be as above. Then any n+1 elements
$f_1, f_2, \ldots, f_{n+1}$ in β have the property that, for almost all $(z_1, \ldots, z_n)$ in
C^n, the set

$$\{\phi(f_{n+1}) \mid \phi \in S, \phi(f_i) = z_i, \quad 1 \leq i \leq n\}$$

is finite.

This is a powerful result. As a simple illustration, take n = 1 and let
X be a piece of one complex-dimensional variety in (z, w)-space C^2. Let β
be the algebra of restrictions of polynomials in z and w to X. We choose a
compact set $K \subset X$ and we take $f_1 = z$, $f_2 = w$. Since the spectrum of β
relative to $\| \|_K$ corresponds to the polynomially convex hull $\hat{K}$ of K in
C^2, the Theorem gives that for almost all a in C, $\{b \mid (a, b) \in \hat{K}\}$ is finite.

The method of proof of the Theorem is highly original and depends on good estimates of $\| F(g_1, \ldots, g_m) \|_K$ where the $g_i \in \beta$ and F is a polynomial in m variables.

Bishop gave two applications of the method in [5]. Let X be a separable complex manifold spread over $\mathbb{C}^n$ such that the analytic functions separate points on X, thus, in particular, X may be a domain in $\mathbb{C}^n$. Let $E(X)$ denote the envelope of holomorphy of X, i.e. the largest domain, spread over $\mathbb{C}^n$, to which every function analytic on X can be analytically extended. Bishop showed

COROLLARY 1: $E(X)$ is holomorphically convex, hence is a Stein manifold.

This result is originally due to Cartan, Thullen, and Oka (see also Gunning and Rossi [10]).

The other application of the method is to the case $n = 1$, i.e. X is a Riemann surface and β is an algebra of holomorphic functions on X which is not identically constant on any component of X. Let K be a finite union of disjoint analytic curves on X. What can be said about the spectrum of β relative to $\| \ \|_K$? For instance, if X is an annulus in the z-plane, K is a concentric circle lying on X, and β is the algebra of polynomials in z restricted to X, it is easy to see that the spectrum of β relative to $\| \ \|_K$ can be identified with the disk which is bounded by the circle K. In general, one has

COROLLARY 2: Let S denote the spectrum of β relative to $\| \ \|_K$. Then $S \setminus K$ can be given the structure of a Riemann surface, with certain discrete identifications, so that the functions in β are holomorphic on $S \setminus K$.

Corollary 2 generalizes results of the author in [22] on the polynomially convex hull of a real analytic curve in $\mathbb{C}^n$. This problem had earlier been treated by Bishop in [6] (see below), and also by Royden in [26].

4. BANACH ALGEBRAS [6]

Let A be a commutative semi-simple Banach algebra with identity. Let X denote the Silov boundary of A and M the maximal ideal space of A. Bishop studied the question of finding conditions under which portions of $M \setminus X$ can be endowed with analytic structure in such a way that the Gelfand transforms of the elements of A are analytic on this structure. His contributions to this problem in [5] and [6], including various important technical Lemmas proved there, have led to research in two directions: first, the problem of uniform approximation in $\mathbb{C}^n$, by polynomials in the coordinates, on curves and Jordan arcs under various regularity conditions. The case of C^1-smoothness was resolved by Stolzenberg in [23], the rectifiable case by H. Alexander [1], and related questions by J.-E. Björk [5], Gamelin [24], and by others.

The other direction was the study of analytic structure associated with function algebras under abstract conditions. A basic result for this work is the following theorem implicit in Bishop's paper [5] (see also [21], Chapter 11):

THEOREM 3: Let A, M, and X be as above and fix an element f in A. (We identify f with its Gelfand transform, so that f is defined and continuous on M.) Choose a connected component W of $\mathbb{C} \setminus f(X)$ and assume that $\exists$ a set of positive plane measure of points $\lambda \in W$ such that $\{p \in M \,|\, f(p) = \lambda\}$ is a finite set. Then the set $f^{-1}(W) = \{p \in M \,|\, f(p) \in W\}$ can be given the structure of a Riemann surface with a certain discrete set of identifications such that the Gelfand transform of each element of A is holomorphic on $f^{-1}(W)$.

It has turned out that versions of this result are true in considerably greater generality. Furthermore, there are close and surprising connections between this result and potential theory as well as with the spectral theory of bounded linear operators on Banach space. These relationships have been pursued in the work of R. Basener [8], B. Aupetit [2], the author and Aupetit [4], and Slodkowski [16].

Also, it has become clear that the classical study of singularity sets of analytic functions of several complex variables, due to F. Hartogs [11], K. Oka [15], and T. Nishino [14], is intimately related to these questions. See [27], [28]. This relationship has led to the creation of a unifying theory of "analytic multivalued functions" (see Slodkowski [16], Aupetit [2]).

Finally, higher-dimensional analogues of Theorem 3 were found by R. Basener [9] and N. Sibony [25], and the study of the higher dimensional case was recently advanced by Slodkowski [17].

5. REMOVABLE SINGULARITIES OF ANALYTIC SETS

In the paper [7] Bishop studied the following questions:

i)   Let $\{A_i\}$ be a sequence of k-dimensional analytic sets in an open subset U of $\mathbb{C}^n$ converging to a set $A \subset U$. Under what conditions is the limit set A again an analytic set in U?

ii)   Let U be as above and let P be an analytic subset of U. Consider a k-dimensional analytic subset A of U \ P. Under what conditions is A ∩ U an analytic subset of U? In other words, when is the singularity set of A on P removable?

Earlier work on these questions had been done by Stoll [18] and by Remmert and Stein in [29].

Using ideas and results from his work in [3] and [5], and in particular the existence of Jensen measures proved in [5] (for a discussion of Jensen measures, see Glicksberg's article in this volume), as well as by constructing new geometric arguments, Bishop in [7] gave several answers to these questions. He showed

THEOREM 4: With $\{A_i\}$ as in Question (i), assume that the 2k-dimensional volumes of the sets A_i are finite and stay bounded as $i \to \infty$. Then A is an analytic subset of U.

THEOREM 5: With A, P, U as in Question (ii), assume that A has finite 2k-dimensional volume. Then $\overline{A} \cap U$ is an analytic subset of U.

Work involving these results has been done in recent years by Shiffman [30], Siu [31], and others.

6. COMMENTS BY HUGO ROSSI

The concept of "special analytic polyhedron" has had tremendous application in the analytic geometry of Stein manifolds. Theorems on extension of holomorphic or meromorphic functions or varieties from the boundary, new and easy proofs of the Remmert-Stein extension theorem for varieties and the Proper Mapping theorem are based on this construction and its conception by Bishop. All these arguments appear in Gunning-Rossi [10], but they are all due explicitly or implicitly to Bishop.

Bishop has a paper [8], on Theorems A and B for Frechet-space valued functions which has had applications in the development of integral formulas (Gleason...), and deformation theory (Douady...).

The paper on "Interpolation of semi-norms," [5], gives a criterion on two sequences to be the Taylor expansions at different points of analytic continuations of each other (in addition to the results mentioned above in Sections 3 and 4).

CITED PAPERS BY ERRETT BISHOP

[1]   Mappings of partially analytic spaces, Amer. J. of Math., vol 83, #2
       (1961), pp. 209-242.

[2]   Some global problems in the theory of functions of several complex
       variables, Amer. J. of Math., vol. 83, #3 (1961), pp. 479-498.

[3]   Partially analytic spaces, Amer. J. of Math., vol. 83, #4 (1961),
       pp. 669-692.

[4]   Differentiable manifolds in complex Euclidean space, Duke Math. J.
       32, #1 (1965), pp. 1-22.

[5]   Holomorphic completions, analytic continuations and the interpolation of
       semi-norms, Ann. of Math., vol. 78, #3 (1963), pp. 468-500.

[6]   Analyticity in certain Banach algebras, Trans. Amer. Math. Soc.,
       vol. 102, #3 (1962), pp. 507-544.

[7]   Conditions for the analyticity of certain sets, Mich. Math. Jour.,
       vol. 11 (1964), pp. 298-304.

[8]   Analytic functions with values in a Frechet space, Pac. Jour. of Math.,
       vol. 12, #4 (1962), pp. 1177-1192.

REFERENCES

[1]   H. Alexander, Polynomial approximation and hulls in sets of finite
       linear measure in C^n , Amer. J. of Math. 93 (1971), pp. 65-74.

[2]   B. Aupetit, Analytic multivalued functions in Banach algebras and uni-
       form algebras, Adv. in Math. 44, #1 (1982), pp. 18-60.

[3]   A. Andreotti and T. Frankel, The Lefschetz theorem on hyperplane
       sections, Ann. Math. 69 (1959), pp. 713-717.

[4]   B. Aupetit and J. Wermer, Capacity and uniform algebras, J. Funct.
       Anal., 28 (1978), pp. 386-400.

[5]   J.-E. Björk, Analytic structures, Summer gathering on Function
       Algebras, Aarhus 1969, pp. 19-28 (Matematisk Inst., Aarhus Univ.).

[6]   E. Bedford and B. Gaveau, Envelopes of holomorphy of certain 2-
       spheres in C^2, Amer. Jour. of Math., vol. 104, #6 (1981).

[7]   A. Browder, Cohomology of maximal ideal spaces, Bull. Amer. Math.
       Soc., 67 (1961), pp. 515-516.

[8]   R. Basener, A condition for analytic structure, Proc. Amer. Math.
       Soc., 36 (1972), pp. 156-160.

[9]   R. Basener, A generalized Silov boundary and analytic structure,
       Proc. Amer. Math. Soc., 47 (1975), pp. 98-104.

[10] R. Gunning and H. Rossi, Analytic Functions of Several Complex
      Variables, Prentice-Hall, Englewood Cliffs, N.J. (1965).

[11] F. Hartogs, Über die aus den singulären Stellen einer analytischen
      Funktion mehrerer Veränderlichen bestehenden Gebilde, Acta Math.
      32 (1909), pp. 57-79.

[12] H. Lewy, On hulls of holomorphy, Comm. Pure and Appl. Math., 13
      (1960), pp. 587-591.

[13] R. Narasimhan, Imbedding of holomorphically complete complex
      spaces, Amer. J. Math., 82 (1960), pp. 917-934.

[14] T. Nishino, Sur les ensembles pseudoconcaves, J. Math. of Kyoto Univ.
      (1962), pp. 225-245.

[15] K. Oka, Note sur les familles de fonctions analytiques multiformes,
      etc., Jour. of Sci. of the Hiroshima Univ. (1934), pp. 93-98.

[16] Z. Slodkowski, Analytic set-valued functions and spectra, Math. Ann.,
      256 (1981), pp. 363-386.

[17] Z. Slodkowski, Lecture at Rickart Symposium, Yale University (June
      1983), to appear.

[18] W. Stoll, The growth of the area of a transcendental analytic set of
      dimension one, Math. Zeitschr., 81 (1963), pp. 76-98.

[19] S. Webster

[20] R. O. Wells and L. R. Hunt, The envelope of holomorphy of a two-
      manifold in C^2, Rice University Studies, vol. 56, #2 (1970), pp. 51-62.

[21] J. Wermer, Banach algebras and several complex variables, Springer-
      Verlag, Graduate Texts in Mathematics, second edition, #35 (1976).

[22] J. Wermer, The hull of a curve in C^n, Ann. Math., 68 (1958),
      pp. 550-561.

[23] G. Stolzenberg, Uniform approximation on smooth curves, Acta Math.,
      115 (1966), pp. 185-198.

[24] T. W. Gamelin, Polynomial approximation on thin sets, Symposium on
      Several Complex Variables, Park City, Utah, 1970, Lecture Notes in
      Mathematics #184, Springer-Verlag (1971), pp. 50-78.

[25] N. Sibony, Analytic structure in the spectrum of a uniform algebra,
      Spaces of Analytic Functions, Kristiansand, Norway, Lecture Notes in
      Mathematics, #512 (1976), pp. 139-165.

[26] H. Royden, Algebras of bounded analytic functions on Riemann surfaces,
      Acta Math., 114 (1965), pp. 113-142.

[27]   J. Wermer, Maximum modulus algebras and singularity sets, Proc. of
       the Royal Soc. of Edinburgh, 86A (1980), pp. 327-331.

[28]   J. Wermer, Potential theory and function algebras, Texas Tech.
       University, Math. Series, Visiting Scholars Lectures 1980, 14 (1981),
       pp. 113-125.

[29]   R. Remmert and K. Stein, Über die wesentlichen Singularitäten
       analytischer Mengen, Math. Ann., 126 (1953), pp. 263-306.

[30]   B. Shiffman, Extending analytic subvarieties, Symposium on Several
       Complex Variables, Park City, Utah, 1970, pp. 208-222, Lecture
       Notes in Math., vol. 184, Springer-Verlag (1971).

[31]   Y.-T. Siu, Techniques of extension of analytic objects, Lecture Notes
       in Pure and Appl. Math., vol. 8, Marcel Dekker, Inc., New York
       (1974) (book).

Contemporary Mathematics
Volume 39, 1985

RECOLLECTIONS

J. L. Kelley

It is a moving experience for me to be here today, to see and talk to old friends, and to remember together. John Wermer has just reviewed Errett's work on several complex variables in the early nineteen sixties, and I understand that Glicksberg, Royden and Stolzenberg will review the work on function algebras and the later contributions to intuitionist analysis. So I will just reminisce, add a little to the narrative of Stefan Warschawski on Errett's career, and speculate a bit on the background of some of Errett's interests.

Last night I looked through a copy of the University's biographical form for Errett, that Murray Rosenblatt got for me. There was no really new information, but I was reminded of many things that had slipped out of my mind. Errett was born in 1928. It struck me, reading the biography, that this was a year of the dragon in the Chinese zodiacal system. So, in Chinese astrology, or mythology, Errett would be considered a dragon -- a rather gentle, reluctant dragon perhaps, but definitely a dragon. He was born in Florence, Kansas, about a hundred miles southeast of the place where I was born a dozen years earlier.

Errett entered the University of Chicago in 1944, when he was sixteen years old. I think, but am not sure, that he entered "the College," which was an experimental college program developed by Robert Hutchins and Mortimer Adler around the "great books." The program admitted promising students without requiring high school graduation, and it attracted many young, exceptionally capable students.

Errett's third and last undergraduate year was 1946-47, and that was my first year at the University of Chicago. I went to Chicago in the summer of 1946, the summer Errett reached eighteen, and I taught a section of calculus and a graduate course on complex function theory. I think Errett was in that summer course on complex functions, and I know he was in my graduate

course in the fall, a course that was called point set topology and was pretty
much about what we now call general topology. I remember him very well.
He looked to be about sixteen, with a sweet self-deprecating smile and a sharp
self-confident mind. He was one of the two or three undergraduates in a
largish class of remarkably able students. Bill Massey, Herman Rubin,
Henry Dye, Matthew Gaffney and four or five other students besides Errett
were very, very good. The last two quarters of the academic year 1946-47 I
taught a course in algebraic topology. I went through the Eilenberg-Steenrod
"Axioms for Homology Theory," which was in manuscript form at the time,
and had a lot of fun, but I remember distinctly that I was disappointed that
Errett was not in the class.

What was the University of Chicago like during 1946, and what kind of
mathematics was happening? Well, the University was recovering from what
was, for Chicago, a relatively low period in the department of mathematics.
Marshall Stone had just signed on as head of the department, and had begun
the build-up, the phenomenal post-war resurgence of the mathematics group
at Chicago. Stone actually arrived in Chicago during 1946-47, Antoni
Zygmund had come earlier, and there were a number of younger faculty. The
most interesting mathematics for me took place around a seminar formally
devoted to functions of positive type, but actually with a broader outlook. It
was quite a seminar. Paul Halmos, Irving Kaplansky, Will Karush, Al
Putnam, Seymour Sherman and W. H. Mayer were regular members, and
Zygmund and Stone came on occasions (occasions when they were talking at
least). The mathematical atmosphere was bracing.

Let me describe the kind of mathematics that was talked about in that
seminar, that was in the air then, because this was the mathematical area of
Errett's early work. The primary objects of study are endowed with a global
algebraic structure as well as a topological or analytic one. The algebra of
continuous real or complex valued functions on a topological space, algebras
of bounded operators on a Hilbert space, topological groups and convolution
algebras of measures on such groups, and the relations between these were
the primary focus. Marshall Stone's study of the $C(X)$ algebra in the nineteen thirties was followed by the brilliant work of Gelfand and the Russian
school: Naimark, Shilov, Raikov, Krein and others. It was a time when new,
exciting relationships were floating up, and we were tantalized by the postwar
delay in getting hold of work done in Russian and in Japan. The fact that the

(norm closed, self adjoint) algebra generated by a normal operator A is iso-morphic to C (spectrum A) cast new light on the spectral theorem, and the true algebraic character of the Fourier transform had only recently been made clear. The Fourier transform became the Gelfand-Fourier representation.

In 1947, after I came to Berkeley from Chicago, I heard that Errett was taking a Master of Science degree. Later, after running into Chicago people at Society meetings, I heard that he was in the Army. This was back in the dark ages when you could actually find someone you wanted to see at the AMS meetings -- before they got so big and crowded you couldn't find anyone or anything. So I kept track of Errett, although I'm not sure that I saw him again until 1954, when he came to Berkeley.

My memory is that he came to Berkeley because he wanted to rework some of the ideas for his dissertation, which was essentially finished but hadn't been written up yet. So he came for a change of scenery or something of the sort, and of course this was postwar, and we were crowded, and he was immediately put to work as an acting instructor. That was the year before he took his Ph.D. degree, in 1954-55. His thesis was written with Paul Halmos on operator theory, and Errett's work thereafter, for some years, was in the heavily worked stream that I've tried to describe.

His academic career was absolutely brilliant. I looked this up last night, and here is what I found: He was an acting instructor in 1954-55, and an instructor for 1955-57 -- that may surprise you, it surprised me a little. But, then I looked up his bibliography and saw that the burst of results began to be published about 1956-57. Whereupon he was immediately an assistant professor. That would have been '57-'58 -- his first year as assistant pro-fessor and '58-'59 was his last year as an assistant professor. He became an associate professor in 1959, and he spent three full years in the rank. How-ever, he went essentially to the top of the professor scale from associate pro-fessor in 1962.

The last years of the nineteen fifties, and the first of the sixties were times of immense productivity on the part of Errett. The papers reviewed by John Wermer and the work Irving Glicksberg discussed all came from this period. Errett was producing beautiful, beautiful mathematics. It was a particularly nice period for Berkeley mathematics in the direction I've de-scribed earlier. Every summer there were new and interesting people to talk to. Most of the people in this room spent at least a month or so in Berkeley

during one of those summers. It was a golden period, and Errett and his
mathematics influenced us all.

I went to India in 1964 and stayed there for a year and a half, so I was
away from Berkeley at the end of Errett's time there. The year 1964 was
the year of the free speech movement, and the anti-war movement was developing momentum. Errett resigned his Berkeley appointment before I knew he
was considering moving. His appointment at UCSD was a remarkably fine
stroke by Professor Warschawski, but not one that I enjoyed.

I have wondered about Errett's interest in the foundations of mathematics. It seems unlikely that this came from his Chicago training, and it
probably was rooted in some of the mathematical activity in Berkeley in the
nineteen fifties. You must remember that the task of formal axiomatization
of mathematics within set theory was effectively accomplished quite late --
I think in the nineteen twenties and thirties, and Gödel's book on the consistency of the continuum hypothesis was published only around 1940.

Few mathematicians knew much about the formalization of mathematics
and its foundations. Indeed, there was some hostility to investigations of the
foundations -- I think it was Murray Protter who remarked that discussing the
foundations of someone's mathematics was like questioning his taste. It was a
time of change, but change was slow. It was 1943 before I even heard of a
function as a set of ordered pairs, and it was not until 1950 that I was reason-
ably sure that I could do mathematics within a formal system of set theory
(and this was certainly because of the Berkeley school).

Alfred Tarski came to Berkeley before the war, A. P. Morse was influenced by him almost immediately, and in the immediate post-war era an
extraordinary center of metamathematics, logic and set theory grew up
around Tarski and the Berkeley school. Remarkable students; Julia Robinson,
C. C. Chang, Sol Feferman and Dana Scott come to mind. And there were
exciting visitors from the Warsaw school and later from the Hebrew University. A lot was going on, and one can well understand Errett's interest.

Some of the work in foundations connected directly with Errett's area of
interest. There was a flurry of feverish activity centered on Boolean algebras
and their structure, and these concern functional analysts as well as logicians
(the class of projections in the spectral resolution of a normal operator is a
nice Boolean algebra that bridges two worlds).

Nowadays the prejudice against the formalization is mostly gone --
perhaps because computers only obey intelligible instructions -- but the
foundations seem to be made of sand. Within the last dozen years, in the
midst of perfectly honorable, reasonable sounding functional analysis -- or in
general topology -- one runs into undecidable propositions. There are propositions that sound reasonable, and even likely, that are known to be consistent with the axioms of set theory, but whose negations are also consistent.
This is not a happy situation, and one has the feeling that it's time to redo
the axioms of mathematics.

But there are other perhaps more profound difficulties. Perhaps the
very character of mathematics needs to change. Errett thought so, and he
was not alone. Here is a related viewpoint: Alex Chorin is an applied mathematician at Berkeley who uses large scale high speed computing to study
mathematical questions. Talking casually in the elevator one day he remarked
on a certain partial differential equation. He said that there is an existence
proof for the equation, a proof that a solution exists. But all known methods
of computation fail; one cannot compute a solution numerically. "It's not
clear what an existence proof means in a case like this," he said.

After Errett moved to UCSD I saw him rather seldom, and we had very
little chance to discuss mathematics. I went on studying the sort of abstract
mathematics that may very well have nothing to do with "real mathematics"
in the sense of the constructivists, and surely has very little to do with
reality. But it's pretty, and I comfort myself by thinking that Matisse's late
work has little connection with reality.

The last time I saw Errett was in San Diego, perhaps a dozen years ago.
I gave a talk on a question in algebraic topology at San Diego State, Errett
came to the lecture, and afterwards we had a little time together, before he
rode off on his motorcycle.

*   *   *

Today, six months after the conference in Errett's honor, as I am
finishing the editing of my remarks, a new book has arrived in the mail. It
is written by Professor M. Hasumi, and the penultimate paragraph of the
Preface begins: "My interest in the subject treated here was first aroused
while I was visiting the University of California at Berkeley in 1962-64. For
this I am indebted to Professors E. Bishop, ... " ..

Contemporary Mathematics
Volume 39, 1985

ASPECTS OF CONSTRUCTIVE ANALYSIS

Halsey L. Royden

I want to talk on some aspects of Errett's work on constructive analysis. When Errett and I were students there was more general interest among mathematicians about questions of foundations and meaning in mathematics. With the advent, early in this century, of set theory with its paradoxes and counter-intuitive examples, many mathematicians believed something must be done to put mathematics on a secure basis. Hilbert and the formalists attempted a program to do this by trying to establish unassailable proofs of consistency for large formal systems in mathematics, thus guaranteeing their freedom from contradiction. A number of other mathematicians took a constructivist position, holding that any proof of the existence of an object, such as a real number, must be one that gives a means by which the object can be constructed or calculated, at least in principle.

Although Poincaré, Weyl, and many others, had constructivist inclinations, the leading protagonist of the constructivist cause was the Dutch mathematician L. E. J. Brouwer. He held that all constructions and proofs should have intuitive meaning, and observed that the requirement that existence proofs be constructive entailed consequences that seemed strange to the classical mathematician. One such consequence was the disallowance of the law of the excluded middle as a tool of mathematical proof; i.e., we are not allowed to assert a priori of a proposition A that either A is true or not A is true. Furthermore, we are generally unable to tell whether two given real numbers are equal or not. The program of Brouwer and his fellow Intuitionists held that a number of classical theorems were untenable. Among these were the theorem that every continuous function on a closed interval had a maximum, the intermediate value theorem, and even the Brouwer fixed-point theorem!

In order to remedy some of the perceived shortcomings of constructivist mathematics, Brouwer and the Intuitionists introduced methods not generally acceptable to most mathematicians and "proved" theorems that were not true in classical mathematics, e.g., that an everywhere defined function on $[0, 1]$ is uniformly continuous. (Errett sometimes referred to these methods as "heuristic" and sometimes as "mystical.")

Positions were strongly held, and arguments were vociferous and polemical in the early decades of this century. By midcentury the arguments had receded, but there was still general interest in foundational questions. One of my teachers, Harold Davenport, told me he had looked into Intuitionism and found it philosophically appealing, but that it was unfortunately difficult or impossible to do the sort of analysis one was used to in the confines of such a system. He gave the example of a theorem in number theory (I believe it was Littlewood's theorem on the alternating abundance of primes of the form $4k+1$ and those of the form $4k+3$) which had originally been proved under the assumption that the Riemann hypothesis was true, and for which a later proof had been given starting from the assumption that the Riemann hypothesis was false. What then was the status of the theorem in question? It was true if the Riemann hypothesis were true or false, but it isn't known whether the Riemann hypothesis is true or false. Fortunately for the number theorist's peace of mind, a later proof was found independent of the Riemann hypothesis.

I do not know of any significant classical theorem whose statement is meaningful constructively but for which we do not have a constructive proof, or at least a clear indication of the lines such a proof would take. We can, however, give somewhat artificial examples, such as the following, to show the nature of a classical proof that cannot be made constructive. Let $e_k = 0$ if $a^n + b^n = c^n$ for all positive integers with $n > 2$ and a, b, c and $n \leq k$, and set $e_k = 1$ otherwise. Then $\Sigma\, e_k\, 2^{-k}$ defines a real number x. If Fermat's last theorem is true, then $x = 0$. If Fermat's last theorem is false, then $x = 2^{-k+1}$, where k is the least value of k for which it is false with a, b, c, $n \leq k$. Thus we have shown that x is a rational number if Fermat's last theorem is false. Can we now assert that x is a rational number. Classically the answer is yes, since Fermat's last theorem is (classically) either true or false, but from a constructivist point of view we cannot assert that x is rational because we are unable to calculate integers p and q such that $x = p/q$.

Thus by midcentury progress toward a sound basis for analysis was disappointingly small. The Hilbert program has not succeeded in giving a

consistency proof for a system rich enough for arithmetic, let alone analysis.
Most mathematicians with sympathy toward the constructivist position believed,
like Davenport, that what one could do in a thorough constructivist system was
so limited that one would have only a greatly truncated fraction of the mathematics one needs. Kleene expressed a widely held view in 1952 when he
wrote: [1]

> "What kind of mathematics can be built within the
> intuitionistic restrictions? If the existing classical
> mathematics could be rebuilt within these restrictions,
> without too great increase in the labor required and too
> great sacrifice in the results achieved, the problem of
> the [foundations of mathematics] would appear to be
> solved. [Intuitionistic] mathematics employs concepts
> and makes distinctions not found in classical mathe-
> matics; and it is very attractive on its own account. As
> a substitute for classical mathematics it has turned out
> to be less powerful and in many ways more complicated
> to develop. "

It was thus with great excitement and pleasure that I read Errett's book
upon its publication. Here was a systematic treatment of analysis from a
thoroughly constructive viewpoint, eliminating the mysticism of latter-day
Intuitionism, and showing that much of the important material of modern
analysis could be dealt with in such a system by methods not far removed
from the classical treatment. The book gives a clear account of the constructive theorems of Brouwer, but simply sidesteps such arcane and irrelevant
questions as the existence of an everywhere defined function which is not
continuous. [2]

In addition to showing that constructive analysis need not be a severe
truncation of classical analysis, Bishop demonstrates that in many cases the
reformulation of a classical theorem to obtain a constructively valid one not

---

[1] Introduction to Metamathematics, p. 51.

[2] Bishop thought that Brouwer gave good heuristic evidence for their non-
existence. On the other hand, there are good heuristic reasons to assume all
real numbers are recursive in some sense. This would seem to imply the
existence of discontinuous everywhere defined functions. Since these are
extra-mathematical considerations, we are probably on sound ground in believ-
ing each of them in appropriate contexts. Having given up the law of the excluded middle, we are no longer forced to conclude that one or two contradictory statements is false.

only gives new insight into the theorem but often also results in a theorem with
more (classical) content than the original.

I was surprised and disappointed that Errett's book has had as little
effect as it has had on the thinking of most present-day mathematicians. I
know Errett came to feel that most mathematicians were unconcerned about
and oblivious to all philosophical questions about meaning in mathematics.
The old myth that constructive mathematics was too complicated, too outré,
and sacrificed too much of classical mathematics has dominated too long. Too
many generations of mathematicians have become enured to a naive set-
theoretic realism. Thus such a work as Errett's, which shows how much one
can accomplish and which should mark the beginning of a vigorous attempt to
rebuild analysis in a constructive context, comes too late when the goal of a
secure foundation for mathematics is no longer believed in, or if believed,
in memory only.

Although my own views on the foundations of mathematics tend toward
the formalist position with an element of pragmatism, I am not immune to the
appeal of the constructivist position. Inspired by Errett's book, I have often
indulged in constructive mathematics, perhaps with a formalist tendency to
think of it as another formal system (but one with considerable philosophical
appeal). I should like to give a brief account of several of my activities in this
direction and some of the discussions I've had with Errett about them.

The first example is the uniform boundedness theorem. Let τ be a
collection of bounded linear operators from a Banach space X to a normed
linear space Y, and suppose that for each $x \in X$ there is a bound M_x such
that $\|Tx\| \leq M_x$ for all $T \in \tau$. Then the classical uniform boundedness
theorem asserts the existence of a uniform bound M such that $\|Tx\| \leq M\|x\|$
for all T and x. The classical proof is indirect and gives us no way of cal-
culating the bound M. Hence we do not expect the theorem to hold in a constructive theory of analysis. This theorem, however, is one of a number of
classical theorems which, although not themselves constructive, have a constructive contra positive: If τ is a collection of linear operators on X to Y
which is not uniformly bounded, then there is an $x \in X$ such that Tx is not
bounded as T ranges over τ.

A constructive theorem which implies this contrapositive is the
following:

THEOREM: Let T_n be a sequence of normed[3] linear operators on a Banach space X to a normed linear space Y and suppose that $\|T_n\| > n3^n$. Then there is an element $x \in X$ such that $\|T_n x\| > n$.

To prove this theorem we must construct the element x whose existence is asserted. We choose a sequence of elements x_n inductively as follows. We assume $x_1, \ldots, x_{n-1}$ have been chosen and choose x_n, so that $\|Tx_n\| > \frac{3}{4}\|T\| \|x_n\|$ and $\|x_n\| = 3^{-n}$. Let $y = x_1 + \cdots + x_{n-1}$. Then

$$\frac{3}{2}\|T\| \|x_n\| < 2\|x_n\| < \|T(y+x_n)\| + \|T(y-x_n)\| \ .$$

Thus one of the two terms on the right is greater than $\frac{3}{4}\|T\| \|x_n\|$ and, replacing x_n by $-x_n$ if necessary, we have

$$\|T(y+x_n)\| > \frac{3}{4}\|T_n\| \|x_n\| \ .$$

We now define x by setting $x = 4\sum_{k=1}^{\infty} x_k$. Since X is complete, and the series is absolutely convergent, it gives well defined element in X. We have

$$\|T_n x\| = 4\left\|T_n\left(\sum_1^n x_k\right) + T_n\left(\sum_{n+1}^{\infty} x_k\right)\right\|$$

$$\geq 4\left[\left\|T_n\left(\sum_1^{n-1} x_k + x_n\right)\right\| - \|T_n\|\left(\sum_{n+1}^{\infty} \|x_k\|\right)\right]$$

$$> 4\left(\frac{3}{4}\|T_n\| 3^{-n} - \|T_n\|\left(\sum_{n+1}^{\infty} 3^{-k}\right)\right)$$

$$\geq 4\|T_n\| 3^{-n}\left(\frac{3}{4} - \frac{1}{2}\right) = \|T_n\| 3^{-n} > n \ .$$

When I showed this proof to Errett, he asked what is the least order of growth for the sequence $\|T_n\|$ needed in order to construct an x with $\|T_n x\| > n$. One can easily see that $\|T_n\| > a^n$ for any $a > 2$ will suffice, but probably much lower orders of growth suffice.

---

[3] In constructive analysis one must distinguish between the concepts of a bounded linear operator and a normed linear operator. The linear operators in the version here are assumed to be normed for the sake of simplicity. The case of bounded operators can be treated similarly, but the technical details are more complicated.

I told Errett that I discovered this proof some time before his book appeared and was not thinking of constructive analysis but only trying to understand clearly the meaning of the uniform boundedness theorem. I was told "That's much the same thing." There is a useful precept here: Even if one is not primarily interested in constructive methods, proofs that are fairly constructive usually give more insight than indirect ones.

Another example where a constructive version gives new insight is the Jordan curve theorem. When asked if it was possible to give a constructive proof of the theorem that the image of a one-to-one map of the unit circle into the plane divides the plane into two regions, Errett, with his usual insight, immediately pointed out the key consideration: "It depends on what you mean by a Jordan curve. You must know constructively that the map is one-to-one; thus you probably need to use the modulus of continuity of the inverse map as part of your data."

This does turn out to be the key. In thinking this over I was able to come up with constructive versions of the separation theorems in the plane using a modulus of "univalency" that can be derived from the modulus of continuity of the inverse mapping. A constructive proof of the Jordan curve theorem has also been given by Berg, Julian, Mines, and Richmann[4] who show by a careful analysis that the classical proofs using the degree of a mapping can be used to give results similar to those stated here. Since my methods are somewhat different, I would like to outline them here.

Let f be a continuous map of the circle (or the unit interval) into the plane. Given $\epsilon > 0$ and $\delta > 0$, we say that the image J of F has the property $P(\epsilon, \delta)$ if, whenever $|f(t_1) - f(t_2)| < \delta$, then one of the segments (or the segment) joining t_1 to t_2 has an image with diameter less than ϵ. Note that if f has modulus of continuity ω, and its inverse has modulus of continuity ω_1, then J has the property $P(\epsilon, \omega \circ \omega_1(\epsilon))$ for every $\epsilon > 0$. This property behaves nicely under uniform approximation: If $|f-g| \leq \eta$, and if the image of f has the property $P(\epsilon, \delta)$ then the image of g has the property $P(\epsilon + 2\eta, \delta - 2\eta)$.

We can now state some constructive separation theorems:

---

[4]The Constructive Jordan Curve Theorem, Rocky Mountain J. of Math. (1975), pp. 225-236.

THEOREM 1: Let J be the image of $[0, 1]$ by a continuous map into the plane, and suppose the J has the property $P(\epsilon, \delta)$. Then any two points x_1 and x_2 in the plane with $\rho(x_i, J) > 2\epsilon$ may be joined by a polygonal arc C such that $\rho(C, J) > \delta/3$.

Observe that in this version we do not require J to be a Jordan arc, i.e., we do not require the map f to be one-to-one. The property $P(\epsilon, \delta)$ asserts, in a sense, that J is within (ϵ, δ) of being one-to-one. That is the key to the proof of this theorem and the following ones: One first establishes (by rather tedious combinatorial arguments) the standard separation theorems for polygons in a rectangular grating and then uses them to establish this and the theorems below in the cases when f is a piecewise linear map into the lines of a rectangular grating. Then the theorems follow immediately from the grating theorems, since we can approximate an arbitrary map by a grating map, and the hypotheses and conclusions of the theorems behave nicely under uniform approximations.

This theorem can be considered a constructive version of the theorem that a Jordan arc doesn't separate the plane, since a Jordan arc has the property $P(\epsilon, \delta)$ for each $\epsilon > 0$ (and some $\delta > 0$ depending on ϵ). Similar versions of the two halves of the Jordan curve theorem are the following:

THEOREM 2: Let J be the image of the unit circle by a continuous map into the plane, and suppose J has the property $P(\epsilon, \delta)$. Then given any three points x_1, x_2, x_3 in the plane with $\rho(x_i, J) > 2\epsilon$, there are two of them which can be joined by a polygonal arc C with $\rho(C, J) > \delta/3$.

THEOREM 3: Let J be the image of the unit circle by a continuous map into the plane, suppose that the diameter of J is d and that J has the property $P(d/3, \delta)$ for some δ. Then there is a point y_o in the plane with $\rho(y_o, J) > \delta/3$ such that if C is any curve connecting y_o to ∞ and J' is any curve whose defining map is uniformly within $\delta/9$ of J, then $\rho(J', C) = 0$.

The last area of constructive analysis that I wish to comment on is the theory of functions of a complex variable. One tends to think that the theorems of this field are constructive if any parts of analysis are. The chapter devoted to this in Bishop's book has a rough unpolished look. Presumably, Errett was only interested in showing that the major theorems of the theory could be proved in the setting of constructive analysis. Errett and I speculated on the possibility of writing a treatment of the subject that was satisfactory to the constructivist mathematician but which the classical mathematician would not perceive to be different from a standard classical text. One cannot, of course, do quite this much, since one will always have to take note of the fact that, constructively, we cannot always tell whether two complex numbers are equal or not.

Since presenting this lecture, I have carried out (partly with Stolzenberg) some development along these lines. One fundamental problem not addressed in Errett's book is that of showing that a holomorphic function f not identically zero on $|z| \leq 1$, is bounded away from zero on some circle of radius close to 1. Classically, this follows from the fact that the zeros of f in $|z| \leq b < 1$ are isolated and hence finite in number. This, of course, is not a method available to us constructively. However, the following theorem has a constructive proof using the Schwarz lemma.

THEOREM: Let f be holomorphic on $|z| \leq 1$ with $|f(z)| \leq M$ there, and suppose $|f(z_o)| \geq \alpha > 0$ for some $|z_o| < \beta < 1$.
Then given η with $\beta \leq \eta < 1$ and $\delta > 0$, there is an integer N depending only on (M, α, β, η) and an $\epsilon > 0$ depending on $(M, \alpha, \beta, \eta, \delta)$ such that for any N points $z_1, \ldots, z_n$ in $|z| \leq \eta$ with $|z_i - z_j| \geq \delta$ we have at least z_k for which $|f(z_k)| > \epsilon$.
COROLLARY: If $C_1, \ldots, C_N$ are N disjoint compact sets in $|z| \leq \eta$ (i.e., with $\rho(C_i, C_j) > 0$ for $i = j$), where N is that of the theorem, then there is at least one of them on which $|f(z)|$ is bounded away from zero.

We thus see that given a function f holomorphic and not identically zero in $|z| \leq 1$, there is a circle $|z| = \eta$ (as close as we wish to $|z| = 1$) on which $|f(z)|$ is bounded from below. Let n be the winding number of f around $|z| = \eta$. Then using a few standard theorems (the Cauchy formula, Liouville's theorem, Riemann's removable discontinuity theorem, etc.) one can find a monic polynomial p(z), all of whose zeros lie in $|z| < \eta$ and a holomorphic function h on $|z| \leq \eta$ such that on $|z| \leq \eta$ we have

$$f(z) = p(z) e^{h(z)}.$$

The function h and the coefficients of p are all calculable by integrals around $|z| = \eta$ involving f'/f.

Contemporary Mathematics
Volume 39, 1985

THE WORK OF ERRETT BISHOP AND UNIFORM ALGEBRAS

I. Glicksberg

It is remarkable that it was only some eight years that Errett Bishop
worked on uniform algebras, for so many basic and truly notable ideas re-
sulted that it would be difficult to write anything on the subject today without
using some of his results or their offspring. My intention here is to try
briefly to indicate the direction and scope of those results, sketch some of the
main ones, and suggest their subsequent influence.

In the fifties, in a relative calm after the initial burgeoning of general
Banach algebra theory, there was a shift to an intensive investigation of their
simplest form, the uniform (or sup-normed) algebra, spurred along by a rush
of attractive ideas and results from (among others) Arens and Singer, Gleason,
Hoffman, and Wermer. Beyond their simplicity of course lay the intimate
connection of uniform algebras with approximation, and it was this connection
which marked Bishop's entry to the field in the late fifties.

1. APPROXIMATION AND ORTHOGONAL MEASURES. Recall that a uniform
algebra A is simply a closed subalgebra of $C(X)$, the supremum normed
space of complex continuous functions on a compact Hausdorff space X, taken
as separating the points of X and containing the constants. Aside from $C(X)$
itself, it is not conjugate closed, and the standard first example is the disc
algebra: for the disc $D = \{z \in \mathbb{C}: |z| \leq 1\}$,

$$A(D) = \{f \in C(D): f \text{ is analytic on } D^o\}.$$

Of course, X can be taken as any closed set between ∂D and D by the maximum principle, and $A(D)$ can also be viewed as $P(D)$, the uniform limits on
D of polynomials in z, or indeed $R(D)$, those of rational functions (in $C(D)$,
so with poles off D). More generally we can replace D by any compact
$K \subset \mathbb{C}$, with X any compact set between ∂K and K, and the fundamental

question of rational approximation in the plane is when equality obtains in the obvious containment

(1) $R(K) \subset A(K)$.

In 1952 Mergelyan [M] showed equality holds in the particular case that $\mathbb{C} \setminus K$ is connected, where Runge's theorem shows $P(K) = R(K)$ as well. (Indeed this last equality trivially implies $\mathbb{C} \setminus K$ is connected, or, equivalently, that K is polynomically convex, so Mergelyan's result gives a complete answer to when $P(K) = A(K)$.)

Bishop's first goal in this area [B2] was to extend this result to a Riemann surface S, replacing polynomials by an appropriate algebra of analytic functions R on S. A very simplified version of one of his results could be stated as follows (where for brevity I have taken his "singular sets" as empty).

THEOREM. Suppose R separates the points in any compact subset of S, and for each point of S contains an element $1 - 1$ nearby. Then for any compact $K \subset S$ which coincides with its R-convex hull,

$$\{s \in S: |f(s)| \leqq \sup |f(K)|, \ f \in R\} ,$$

one has the uniform closure $(R|K)^- = A(K)$.

Most of his methods, which included use of Mergelyan's result, were classical, but the basic step was functional analytic: by the Hahn-Banach and Riesz representation theorems one has only to show a measure of orthogonal to the smaller algebra $(R|K)^-$ is orthogonal to the larger. In the course of the proof one highly significant fact concerning rational approximation in C is proved, namely

BISHOP'S SPLITTING LEMMA. Any measure $\mu \perp R(K)$ is the sum of finitely many orthogonal measures with arbitrarily small support.

This immediately shows that in general equality in (1) is purely local, as was recorded later by Garnett [Ga]; he observes, for example, that if $\mathbb{C} \setminus K$ has components all of diameter $\geqq 4\delta > 0$ then since any closed disc $D(z, \delta)$ of radius δ meets K in a set whose complement is connected we have $\mu \perp R(K)$ a sum of measures orthogonal to $f \in A(K)$, since f restricts to an element of each algebra $A(K \cap D(z, \delta)) = P(K \cap D(z, \delta))$ (by Mergelyan). Thus, equality in (1) obtains here. (Vitushkin's general approach [V], of course, also shows the local character of the question, but the route through Bishop's lemma is far more direct and much simpler.)

This approach through duality led Bishop naturally to consider the form of the measures orthogonal to $R(K)$ carried by ∂K (the Šilov boundary for $R(K)$ and $A(K)$). In the classical case of the disc these are entirely charac- terized by the F. and M. Riesz theorems, and Bishop next attempted to ex- tend the latter to the setting of Mergelyan's theorem, i.e., for K polynomi- ally convex. In [B4] he treated the case in which K has connected interior U, with the basic idea to obtain the general measure μ on ∂K orthogonal to $R(K)$ as an "analytic differential," which, since $\partial U \subset \partial K$ is rarely of finite length, was taken as the w^* limit of the measures provided by $g(z)\,dz$ (with g analytic on U) on a sequence of simple closed rectifiable curves γ_n in U converging to ∂U. It is easy to see that there is a unique candidate for g, the Cauchy transform $\hat{\mu}$ of μ,

$$(2) \qquad \hat{\mu}(z) = \frac{1}{2\pi i} \int \frac{d\mu(\zeta)}{\zeta - z}$$

since Cauchy's theorem and w^* convergence imply that for $z \in U$ and $n = n_z$ large

$$g(z) = \frac{1}{2\pi u} \int_{\gamma_n} \frac{g(\zeta)\,ds}{\zeta - z} \to \frac{1}{2\pi i} \int \frac{d\mu(\zeta)}{\zeta - z} .$$

(Of course the integral (2) converges absolutely for (area) almost all z as the convolve of μ and a locally integrable function.) Using various classical re- sults, in particular, a result of Carathéodory on Riemann maps, the F. and M. Riesz theorem and Mergelyan's result itself, Bishop indeed showed that μ is this w^* limit, for any sequence $\{\gamma_n\}$. Via the Riemann map of D onto U he thus obtained μ as the image of a measure on ∂D orthogonal to the disc algebra $A(D)$, i.e., of $h(e^{i\theta})d\theta$ for $h \in H_o^1$, so characterizing the ortho- gonal measures. As corollaries he obtained the Lebesgue-Walsh theorem and a version of Wermer's maximality theorem [G1] in this setting.

Naturally "analytic differentials" here are also orthogonal to $A(K)$: for each measure $\hat{\mu}(z)\,dz\big|_{\gamma_n}$ is. Since the role of Mergelyan's theorem in Bishop's proof was confined to showing that if a measure $\nu \perp R(K)$, and so with $\hat{\nu} = 0$ off K, also has $\hat{\nu} = 0$ on U, then $\nu = 0$, only an alternative argu- ment for this step is needed to obtain a proof of Mergelyan's result itself (in this simple case). By the time Bishop had seen how to extend his results of [B4] to the general case (in which K^o has countably many components) in [B11] he had himself provided the necessary alternative argument in his remarkable paper on minimal boundaries [B7]. Thus [B11] includes a new proof of Mergelyan's general result, essentially as an aside; the main result shows each measure on ∂K orthogonal to $R(K)$ is the norm convergent sum of orthogonal measures each carried by the boundary of a component U of the interior of K and indeed an "analytic differential" in U (so $\perp A(K)$). Later, armed with the F. and M. Riesz theorem for Dirichlet algebras provided by the work of Helson and Lowdenslager [HL] and the Hoffman-Wermer modification theorem [G1] which adapted Forelli's approach to that same classical result [F] in its proof, Wermer and I [GliW] were able to remove almost all the classical components of Bishop's proof to get a function algebra proof using essentially only the Lebesgue-Walsh theorem from classical analysis. Further simplification, including a new approach to the last result, was given by Carleson [C2]. Finally, when a generally applicable abstract F. and M. Riesz theorem for function algebras was found [Gli2], the extension of Bishop's decomposition for measures orthogonal to $R(K)$ to the general case led to the surprising fact that equality in (1) is in fact a self-adjoint problem: only the real orthogonal measures have to coincide, or, alternately, $(\mathrm{Re}R(K))^- = (\mathrm{Re}A(K))^-$ suffices.

Two further contributions [B5, 10] to approximation theory should be mentioned at this point, both related to the question: for a thin compact set $K \subset C$ can one, given $f_o, f_1, \ldots, f_n$ in $C(K)$, simultaneously approximate these uniformly by $p, p', p^{(2)}, \ldots p^{(n)}$ with p a polynomial in z? Of course the f_i would be related if K contained rectifiable arcs. But Bishop showed this could be done without restriction with $K = \phi[0, 1]$ a Jordan arc with no rectifiable subarcs, but satisfying a certain Lipschitz-like condition (asserting that for a dense set of parameter values t for points z close to $\phi(t)$, $|\phi(t) - z|^c$, up to a factor, is a lower bound for the distance from z to one of the subarcs $\phi[0, t]$, $\phi[t, 1]$, with c independent of t). The argument again uses orthogonal measures and is very ingenious; to show such arcs exist takes a combination of three constructions and again is very clever. In the later paper [B10], Bishop used distributions to show that, given a compact nowhere dense set $K \subset \mathbb{C}$ and a totally disconnected subset E one can simultaneously approximate f_o on K and $f_1, \ldots, f_n$ on E by $p, p', \ldots p^{(n)}$.

2. PEAK SETS AND REPRESENTING MEASURES. We now arrive at what, for me, are Bishop's most elegant and direct applications of the simplest properties of uniform algebras; they give results which have had enormous influence on the subject.

The well known Šilov boundary ∂ provides a smallest closed subset of X which yields the supremum norm of any element f of our algebra A: $\|f\| = \sup|f(\partial)|$. Bishop next raised the question [B7] of whether there is a (not necessarily closed) smallest subset M with this property, a <u>minimal boundary</u>. It is easy to see this fails to hold in general, but when X is metric the answer is affirmative: there the supremum can easily be achieved at a single "peak point, " which would then necessarily lie in any boundary, and Bishop showed the set of all such points provides the desired M. We need a

DEFINTION. <u>A closed set</u> $F \subset X$ <u>is a peak set for</u> A <u>if some</u> $f \in A$ <u>has</u> $f(F) = 1$, $|f| < 1$ <u>off</u> F, <u>and then</u> f <u>is said to peak on</u> F. (Of course if F is a singleton we shall call it a peak point.)

A peak set is then just a set F where some non-zero $g \in A$ assumes a value z of maximum modulus: for then $f = \dfrac{1}{2z}(g+z)$ peaks on F. Bishop's first main observation is that peaking within a peak set can be converted to peaking over X, his basic

PEAK SET LEMMA. <u>If</u> F <u>is a peak set for</u> A <u>and</u> $F_o \subset F$ <u>is a peak set for</u> $A|F$, F_o <u>is a peak set for</u> A (i.e., if $f \in A$, $f(F_o) = 1$ <u>and</u> $|f| < 1$ <u>on</u> $F\backslash F_o$, <u>there is an</u> $h \in A$ <u>with</u> $h(F_o) = 1$, $|h| < 1$ <u>on</u> $X \backslash F_o$).

We can assume $|f| < 3/2$ on X, replacing f by $(f+r)/(1+r)$ for $r \in \mathbb{R}$ large. Suppose $g(F) = 1$, $|g| < 1$ on $X\backslash F$, and set

$$K_n = \{x \in X: 1 + 2^{-n-1} \leqq |f(x)| \leqq 1 + 2^{-n}\}, \quad n \geqq 1 .$$

Then $K_n \cap F = \emptyset$ and so there is an integer k_n with $|g^{k_n}| < 2^{-2n}$ on K_n. We now set

$$h = \sum_{1}^{\infty} 2^{-n} f g^{k_n} ,$$

which lies in A since the series converges uniformly. Now $h(F_o) = 1$ and $|h(x)| \leqq |f(x)| < 1$ for $x \in F\backslash F_o$, and indeed $|h(x)| < 1$ for any $x \notin F$ with $|f(x)| \leqq 1$ since then $|g(x)| < 1$. But if $|f(x)| > 1$ then $x \in K_m$, some $m \geqq 1$, so

$$|h(x)| \leq f(x) \left[\sum_{n \neq m} 2^{-n} + 2^{-m} \cdot 2^{-2m} \right]$$

$$\leq (1 + 2^{-m})(1 - 2^{-m} + 2^{-3m})$$

$$\leq 1 - 2^{-2m} + 2^{-3m} + 2^{-4m} = 1 - \frac{2^{2m} - 2^m - 1}{2^m} < 1$$

and the proof is complete.

Now if we consider a maximal decreasing chain of peak sets lying in one, F, the lemma shows their intersection must be a singleton $\{x_o\}$. Since this is a G_δ, it is in fact the intersection of a countable set of peak sets, hence trivially a peak set, so x_o is a peak point (and any intersection of peak sets which is a G_δ is a peak set). Thus the set M of all peak points is a boundary, and necessarily lies in any other, hence is our minimal boundary.

That M is in fact nice, indeed, a G_δ, Bishop obtained from a second beautiful lemma, which doesn't require metrizability or indeed an algebra.

BISHOP'S 1/4-3/4 CRITERION. Suppose A is a linear subspace of C(X), x ∈ X, and for each neighborhood U of x there is an f ∈ A of norm 1 with $|f(x)| > 3/4$, $|f(X) \setminus U)| < 1/4$. Then x is an intersection of peak sets for A.

This has the very important consequence that

(3)   any representing measure λ for p ∈ M (i.e., a $\lambda \geq 0$ with $f(p) = \int f d\lambda$, f ∈ A) is a point mass, and that for any p ∉ M there is a representing measure λ with $\lambda\{p\} = 0$.

Finally Bishop applies his minimal boundary to R(K):

(4)   If K is nowhere dense in $\mathbb{C}$ (so M ⊂ ∂K = K) and K\M has zero area then R(K) = C(K)(= A(K) of course). Thus in fact M = K follows.

The last result improved on a classical result of Hartogs and Rosenthal [G1] and seems to mark the real introduction of the Cauchy transform (2) as a tool in rational approximation.[1] From what has been obtained so far his proof is now very simple. Recall that $\hat{\mu}(z)$ exists for a.a.z. and $\hat{\mu} = 0$ a.e. implies $\mu = 0$ (as an easy consequence of Fubini's theorem [G1, p. 46]).

---

[1]The use of orthogonal measures and the Cauchy transform already occur in the earlier work of Beurling.

Of course $\hat{\mu}(z) = 0$ for $z \notin K$ if $\mu \perp R(K)$, and if $z \in K$ and $\hat{\mu}(z)$ exists then of course $\mu\{z\} = 0$; if $\hat{\mu}(z) \neq 0$ as well, then since f rational with poles off K implies the same is true of $g(\zeta) = \dfrac{f(\zeta) - f(z)}{\zeta - z}$ we have

$$0 = \int g d\mu = \int \frac{f(x)\, \mu(d\zeta)}{\zeta - z} - f(z) \int \frac{\mu(d\zeta)}{\zeta - z}$$

$$= \int \frac{f(x)}{\zeta - z} \mu(d\zeta) - f(z)\, \hat{\mu}(z) \quad .$$

Thus $f(z) = \dfrac{1}{\hat{\mu}(z)} \displaystyle\int \dfrac{f(\zeta)\, \mu(d\zeta)}{\zeta - z}$ holds for $f \in R(K)$, and now if f were to peak at z, since $\mu\{z\} = 0$ we would have

$$1 = \lim f^n(z) = \frac{1}{\hat{\mu}(z)} \int \frac{\lim f^n(\zeta)}{\zeta - z} \mu(d\zeta) = 0 \,.$$

So $\hat{\mu}(z) \neq 0$ implies $z \notin M$, and if $K \setminus M$ has zero area $\hat{\mu} = 0$ a.e., $\mu = 0$, and we see $R(K) = C(K)$ follows, as desired.

Thus in this single paper Bishop lays the foundation for most of what we know about peak sets, peak points and their characterization in terms of representing measures, and the systematic use of the Cauchy transform in rational approximation, and in fact, provides a new proof of the existence of the Šilov boundary as well!

Various consequences of these results should be noted. In particular (3), along with some functional analysis, yields the quite useful dual characterization of (intersections of, or generalized) peak sets: [Gli1, G1] a closed set $F \subset X$ is one iff $\mu_F \perp A$ for all $\mu \perp A$. As Bishop notes in [B7], Karel de Leeuw found another proof of (3) via Choquet's theorem [Ch, P] and this clearly led to their joint paper [B9] in which Choquet's metric restrictions are removed. (I will return to this later.)

Finally the argument using the Cauchy transform, beyond providing a generally useful tool, led to Wilken's theorems [Wi] that (for general K) any measure on ∂K orthogonal to $R(K)$ and singular with respect to all representing measures vanishes, and that if $R(K) \neq C(K)$ some Gleason part for $R(K)$ has positive area, thus extending (4); in fact [Wi] shows each non peak point lies in a part of positive area, so there are only countably many non-peak point parts. Further refinements have been given by Browder [Br].

The notion of a part, an equivalence class in the spectrum (or space of multiplicative linear functionals) M of A under the (not at all apparent)

equivalence given by $\varphi \sim \psi$ iff $\|\varphi - \psi\|_{A^*} < 2$, was one of the inviting
ideas introduced by Gleason in 1957 [G1]. Bishop showed two equivalent ele-
ments of M have mutually bounding representing measures, a result which
has proved quite useful. Much more useful and basic was his proof, given as
a lemma in [B21] (discussed here by Wermer) that every multiplicative linear
functional φ on A is represented by a Jensen measure λ, i.e., a probability measure representing φ which also satisfies Jensen's inequality

$$(5) \qquad \log |f(\varphi)| \leq \int \log |f| \, d\lambda, \quad f \in A.$$

Arens [A] had obtained such a measure for a very special algebra, the big
disc algebra he and Singer studied earlier [AS]. It would be hard indeed to
cite their uses to gauge the importance of Jensen measures and stay within
bounds; instead let me just refer you to Gamelin's recent book devoted to
applications of Jensen measures [G2], and point out that their existence
played an essential role in Cole's counterexample to the peak point conjecture
(that if the spectrum M coincides with the set M of peak points then A
would have to be C(M)), while it also can be made the basis of Wermer's subharmonicity theorem [W1], or various extensions of Radó's theorem [CG1,2].

Bishop's proof of the existence of Jensen measures runs as follows. We
consider

$$Q = \left\{ u \in C_R(X): u > \frac{1}{n} \log |f| \text{ for some } f \in A \text{ with } |f(\varphi)| = 1, n \in \mathbb{Z}_+ \right\}.$$

Clearly Q is open, and since $\frac{m}{n} \log |f| = \frac{1}{n} \log |f^m|$ while $\varphi(f^m) = \varphi(f)^m$,
it is invariant under positive rational dilations. Moreover since $u_i > \frac{1}{n_i} \log |f_i|$
for $i = 1, 2$, implies $u_i > \frac{1}{n_1 n_2} \log |f_1^{n_2} f_2^{n_1}|$, we conclude Q is an open
cone. But trivially Q cannot meet the cone of non-positive functions in
$C_R(X)$: for the $0 \geq u > \frac{1}{n} \log |f|$ for some $f \in A$ with $|f(\varphi)| = 1$ while by
compactness $\frac{1}{n} \log |f| \leq -\delta < 0$ or $|f| \leq e^{-\delta} < 1$, so $|f(\varphi)| \leq \|f\| < 1$.

Thus from the separation theorem we have a measure λ yielding a
functional > 0 on Q and ≤ 0 on the non-positive cone, so $\lambda \geq 0$. Normal-
izing λ so that λ is 1 at the element 1 of Q we have a probability measure
λ for which

$$\int (\log (|f| + \epsilon) \, d\lambda \geq 0 \quad \text{if } f \in A \quad \text{has } |f(\varphi)| = 1.$$

By monotone convergence then $\int \log |f| \, d\lambda \geqq 0$ for such f, whence

$$\int \log \frac{|f|}{|f(\varphi)|} \, d\lambda \geqq 0, \quad \text{or} \quad \int \log |f| \, d\lambda \geqq \log |f(\varphi)|$$

whenever $f(\varphi) \neq 0$, so (5) follows. Finally, applying the inequality to $e^{\pm f}$ shows

$$\int \text{Re } f \, d\lambda = \text{Re } f(\varphi), \quad f \in A,$$

whence λ is a representing measure.

Along with the result, we turn to next, this is among the most used standard tool which Bishop provided for uniform algebra theory.

3. ANTISYMMETRY. It is to Šilov that we are indebted for the notion of antisymmetry and the suggestion that an antisymmetric decomposition could be obtained; but it is to Bishop that we owe the result. Šilov called A antisymmetric if it contained no non-constant real valued functions, and so was as far as possible from symmetric. Calling a subset F of X a set of antisymmetry if $A|F$ contained no non-constant real elements, he proposed decomposing X into a collection of sets of antisymmetry from which, via the restrictions of A, one could recover A. His approach was ultimately carried out by Bishop [B12], transfinitely refining successive decompositions provided by the real elements of the restriction algebra.

BISHOP'S GENERALIZED STONE-WEIERSTRASS THEOREM. The set K of maximal sets of antisymmetry forms a closed, pairwise disjoint covering of X, for which

    1) $f \in C(X)$ and $f|K \in (A|K)^{-}$ for all $K \in K$ imply $f \in A$, and

    2) $A|K$ is closed.

Thus in many instances questions can be reduced to the antisymmetric setting where matters may be easier to handle. To mention a few notable cases, Hoffman and Wermer used this approach to obtain their famous result that Re A closed implies $A = C(X)$ [HW], while Stolzenberg [S] (and following him, Alexander [Al]) used it to get very strong results on analytic structure in spectra (based on those of Bishop and Wermer mentioned in Wermer's talk). The utility has been great, but the proof (excluding 2)) can be now given in a paragraph, because of the Krein-Milman theorem, so we shall do so:

Trivially a point is a set of antisymmetry, as is the union of any two such sets which meet. Thus the union of all sets of antisymmetry containing

a given one is another, and necessarily maximal, and closed since taking the
closure preserves the property. So K forms a closed pairwise disjoint
covering. For (1), to see $f \in A$ we need only see f is orthogonal to all
orthogonal measures, hence by Krein-Milman to all extreme elements of
ball A^+. If μ is one, its closed support S is a set of antisymmetry, as we
next note, and thus $S \subset K \in K$ and $f|K \in (A|K)^-$ will imply $\int f d\mu = 0$ as
desired. But for $g \in A$ real on S and, say, with $0 < g < 1$ there,

$$\mu = \|g\mu\| \frac{g\mu}{\|g\mu\|} + \|(1-g)\mu\| \frac{(1-g)\mu}{\|(1-g)\mu\|}$$

expresses μ as a convex combination of two elements of ball A since

$$\|g\mu\| + \|(1-g)\mu\| = \int gd|\mu| + \int (1-g)d|\mu| = \int d|\mu| = 1 ,$$

so g must be constant mod $|\mu|$, thus constant on S as required, and (1) is
proved.

In fact 2) follows from a stronger fact: each maximal set of antisymmetry is an intersection of peak sets (so a peak set if X is metric). Indeed
any closed G_δ set F in X which is a union of elements of K is itself a
peak set, as a consequence of the cited characterization of intersections of
peak sets and the Bishop-de Leeuw-Choquet theorem; the same argument shows
we need only assume $F \cap K$ is an intersection of peak sets for each $K \in K$
(cf. [Gli1, 3.3]).

4. INTERPOLATION. Rudin [R] and Carleson [C] independently showed that
any compact Lebesgue null set $F \subset \partial D$ is an interpolation set for the disc
algebra: $A(D)|F = C(F)$. Bishop [B17] noting that Lebesgue null sets are
null for all orthogonal measures by the F. and M. Riesz theorem, thoroughly
generalized the subject of interpolation by proving that for <u>any closed sub-</u>
<u>space</u> B <u>of</u> $C(X)$, <u>a compact</u> $F \subset X$ <u>for which the restriction</u> $\mu_F = 0$ <u>for</u>
<u>each</u> $\mu \perp B$ <u>is a set of interpolation for</u> B: $B|F = C(F)$. <u>Indeed for each</u>
$f \in C(F)$ <u>and positive</u> $g \in C(X)$ <u>with</u> $|f| < g$ <u>on</u> F, <u>we have a</u> $b \in B$ <u>with</u>
$b|F = f$, $|b| < g$ <u>on</u> X.

(Thus F is what is now called a peak interpolation set: both an intersection of peak sets and an interpolation set. Conversely if B is an algebra
any such set has $\mu_F = 0$ for all $\mu \perp B$.)

Needless to say this soon led to the general dual condition for an inter-
polation set F (that $\|\mu_F\| \leq C \|\mu_{X \setminus F}\|$ for all $\mu \perp B$, C fixed [Gli2]),

and has inspired a variety of related investigations, from interpolation in special uniform algebras (such as $A(K)$ for K in $\mathbb{C}^n$) to algebras of smooth functions.

5. BANACH SPACE RESULTS. Although most of the results described so far originated in the uniform algebra context it is hardly surprising that some involve only the linear structure, as for example, Bishop's $1/4$-$3/4$ criterion, and his interpolation theorem. Another, which arose from his consideration of minimal boundaries as noted earlier is the celebrated Bishop-de Leeuw Theorem.

Recall that if K is a compact convex subset of a locally convex space Y we can assign to any Borel probability measure μ on K a unique element $\int k d\mu(k)$ of K, the resultant of μ (most simply as the unique element k_o of K with $\langle k_o, y^* \rangle = \int \langle k, y^* \rangle d\mu(k)$ for all $y^* \in Y^*$). Choquet [Ch] proved that if K is metrizable each $k_o \in K$ is the resultant of a μ carried by K^e, the set of extreme points of K, improving on the Krein-Milman theorem (which yields a measure on the closure of K^e). The Bishop-de Leeuw result removes the metric restriction, but matters are necessarily more involved: the measure μ is carried by K^e only in the sense that it vanishes on each Baire set in K disjoint from K^e. As indicated in Phelps' book [P] the result has applications in analysis, probability, potential theory, and functional analysis itself; that it has application to uniform algebras beyond those already noted can be seen in the first chapter of the cited book of Gamelin on Jensen measures [G2].

Finally there are the very well known results of Bishop and Phelps [B13, 20] that the support points of a closed convex set C in a Banach space Y (the points where functionals maximize over the set) are dense in its topological boundary, while the functionals which maximize on C are norm dense in Y^* among those which are bounded above on C. There is a host of consequences; we refer the reader to the survey by Ekeland [E].

BIBLIOGRAPHY

[A1]   H. Alexander, Polynomial approximation and analytic structure, <u>Duke Math. J.</u> 38 (1971), 123-135.

[Ar]   R. Arens, The boundary integral of $\log |\varphi|$ for generalized analytic functions, <u>T. A. M. S.</u> 86 (1957), 57-69.

[ArS1] R. Arens and I. Singer, Function values as boundary integral, <u>P. A. M. S.</u> 5 (1954), 735-745.

[ArS2] R. Arens and I. Singer, Generalized analytic functions, <u>T. A. M. S.</u> 81 (1956), 379-393.

[B2]   "Subalgebras of functions on a Riemann surface," <u>Pacific J. Math.</u>, 8, No. 1 (1958), 29-50.

[B3]   "Measures orthogonal to polynomials," <u>Proc. Nat'l. Acad. Sci.</u>, 44, No. 3 (1958), 278-280.

[B4]   "The structure of certain measures," <u>Duke Math. J.</u>, 25, No. 2 (1958), 283-290.

[B5]   "Approximation by a polynomial and its derivatives on certain closed sets," <u>Proc. Amer. Math. Soc.</u>, 9, No. 6 (December 1958), 946-953.

[B7]   "A minimal boundary for function algebras," <u>Pacific J. Math.</u>, 9, No. 3 (1959), 629-642.

[B8]   "Some theorems concerning function algebras," <u>Bull. Amer. Math. Soc.</u>, 65, No. 2 (1959), 77-78.

[B9]   (with Karel de Leeuw) "The representations of linear functionals by measures on sets of extreme points," <u>Annales de L'Institut Fourier</u>, 9 (1959), 305-331.

[B10]   "Simultaneous approximation by a polynomial and its derivatives," <u>Proc. Amer. Math. Soc.</u>, 10, No. 5 (1959), 741-743.

[B11]   "Boundary measures of analytic differentials," <u>Duke Math. J.</u>, 27, No. 3 (1960), 331-340.

[B12]   "A generalization of the Stone-Weierstrass Theorem," <u>Pacific J. Math.</u>, 11, No. 3 (1961), 777-783.

[B13]   (with R. R. Phelps) "A proof that every Banach space is subreflexive," <u>Bull. Amer. Math. Soc.</u>, 67, No. 1 (1961), 97-98.

[B12]   "A general Rudin-Carleson Theorem," <u>Proc. Amer. Math. Soc.</u>, 13, No. 1 (1962), 140-143.

[B20]   (with R. R. Phelps) "The support functionals of a convex set," <u>Proc. Symposia in Pure Math.</u>, VII (1963), 27-35.

[B21]   "Holomorphic completions, analytic continuations and the interpolation of semi-norms," <u>Annals of Math.</u>, 78, No. 3 (1963), 468-500.

[B22]   "Representing measures for points in a uniform algebra," <u>Bull. Amer. Math. Soc.</u>, 70, No. 1 (1964), 121-122.

[B24]   "Uniform algebras," <u>Proc. Conference on Complex Analysis</u>, Minneapolis (1964), 272-280.

[Br]   A. Browder, An Introduction to Function Algebras, Benjamin, N.Y.
          1969.

[C1]   L. Carleson, Representations of continuous functions, Math. Z. 66
          (1957), 447-451.

[C2]   L. Carleson, Mergelyan's theorem on uniform polynomial approxi-
          mation, Math. Scand. 15 (1964), 167-175.

[Ch]   G. Choquet, Existence des représentations integrales au moyen des
          points extrémaux dans les cones convexes, C. R. Acad. Sci., Paris,
          243 (1956), 699-702.

[Co]   B. J. Cole, One point parts and the peak point conjecture. Thesis,
          Yale University, 1968.

[CG1]   B. J. Cole and I. Glicksberg, Jensen measures and a theorem of
          Radó, J. of Funct. Anal., 35 (1980), 26-41.

[CG2]   B. J. Cole and I. Glicksberg, The spread of pointwise bounded con-
          vergence over the non-Šilov spectrum, in preparation.

[E]   I. Ekelund, Non convex minimization problems, B. A. M. S., 1 (1979),
          443-474.

[F]   F. Forelli, Analytic measures, Pac. J. Math., 13 (1963), 571-578.

[G1]   T. W. Gamelin, Uniform Algebras, Prentice Hall, Englewood Cliffs,
          N.Y., 1969.

[G2]   T. W. Gamelin, Uniform Algebras and Jensen Measures, Cambridge
          Univ. Press, Cambridge, 1978.

[Ga]   J. Garnett, On a theorem of Mergelyan, Pac. J. Math., 26 (1968),
          461-467.

[Gl]   A. Gleason, Function algebras, Seminar on Analytic Functions II,
          Inst. for Adv. Study, Princeton (1957), 213-226.

[Gli1]   I. Glicksberg, Measures orthogonal to algebras and sets of antisym-
          metry, T. A. M. S., 105 (1962), 415-435.

[Gli2]   I. Glicksberg, The abstract F. and M. Riesz theorem, J. of Funct.
          Anal., 1 (1967), 109-122.

[GW]   I. Glicksberg and J. Wermer, Measures orthogonal to a Dirichlet
          algebra, Duke Math. J., 30 (1963), 661-666.

[HL]   H. Helson and D. Lowdenslager, Prediction theory and Fourier
          series in several variables I, Acta Math., 99 (1958), 165-202.

[HW]   K. Hoffmann and J. Wermer, A characterization of C(X), Pac. J.
          Math., 12 (1962), 941-944.

[M]   S. N. Mergelyan, Uniform approximation to functions of a complex
          variable, Uspehi Mat. Nauk, 7 (1952), No. 2 (48), 31-122. A. M. S.
          Transl., Ser. 1, Vol. 3, 281-286.

[P]   R. R. Phelps, Lectures on Choquet's Theorem, Van Nostrand,
          Princeton, 1966.

[R]   W. Rudin, Boundary values of continuous analytic functions,
          P. A. M. S., 7 (1956), 808-811.

[St]   G. Stolzenberg, Uniform approximation on smooth curves, Acta
            Math., 155 (1966), 185-198.

[S]   G. W. Šilov, On decomposition of a commutative normed ring in a
            direct sum of ideals, Mat. Sbornik, 32 (1954), 353-364. A.M.S.
            Transl., (2) 1 (1955), 37-48.

[V]   A. G. Vitushkin, Analytic capacity of sets and problems in approxi-
            mation theory, Uspehi Mat. Nauk, 22 (1967), 141-199. Russian
            Math. Survey, 22 (1967), 139-200.

[Wi]   D. Wilken, Lebesgue measure for parts of R(X), P.A.M.S. 18
            (1967), 508-512.

[W1]   J. Wermer, Subharmonicity and hulls, Pac. J. Math., 58 (1975),
            283-290.

[W2]   J. Wermer, Banach Algebras and Several Complex Variables,
            Springer, N.Y., 1976.

[W3]   J. Wermer, The work of Errett Bishop and several complex
            variables.

Contemporary Mathematics
Volume **39**, 1985

REMEMBRANCES OF ERRETT BISHOP

Anil Nerode [N], George Metakides [M] and Robert Constable [C]

[N]   After the publication of his book <u>Constructive Analysis</u>, Bishop made a
tour of the eastern universities that included Cornell. He told me then that he
was trying to communicate his viewpoint directly to the mathematical com-
munity, rather than through the logicians. He associated the logicians with
defending the turf of codified formal systems, while he himself believed in the
free exercise of positive affirmative mathematical faculties, free of artificial
formal limitations. After the eastern tour was over, he said the trip may
have been counterproductive. He felt that his mathematical audiences were
not taking the work seriously. He was surprised to get a more sympathetic
hearing from the logicians. I told him that I thought that the general reaction
of mathematicians to his work, like the reaction to Brouwer and Heyting before him, was mainly due to his using familiar words (like function) in an unfamiliar way. The mathematical listener hearing him for only an hour could
not grasp his meaning. He granted this possibility, but thought that his meanings, not the meanings used by the classically trained mathematicians, were
the natural ones.

The particular lecture Bishop gave at Cornell was on possible constructive versions of the Riemann mapping theorem. As long as he stuck to the
positive affirmative part (that is, the constructive proof with constructive
hypotheses about the shape of the region), the audience seemed happy. He lost
the audience when he gave a Brouwerian counterexample to an uncritical constructive rendering of the Riemann mapping theorem, without a strong restriction on the shape of the boundary. (This counterexample involved two circles
with a narrow tube connecting them, and the usual Brouwerian argument that if
there were a constructive version of the full theorem without special constructive constraints on the boundary, then some famous problem like Goldbach's
conjecture would be decided.) I mentioned in the question period at the end of

the lecture that the Brouwerian counterexample could be reformulated in classical terms as follows: Any mapping was necessarily discontinuous which assigned to the data of the problem (including the shape of the region) a value which is a Riemann mapping function. I also said (which is true) that this would certainly lead to construction of a recursive region of the desired sort, having no recursive Riemann mapping function.

After the lecture he mentioned tribulations in the reviewing process when he submitted the book for publication. He mentioned that one of the referee's reports said explicitly that it was a disservice to mathematics to contemplate publication of this book. He could not understand, and was hurt by, such a lack of appreciation of his ideas. One of the reasons for the lecture tour was to be sure his ideas got a hearing.

In the next dozen years his students and disciples had a hard time developing their careers. When they submitted papers developing parts of mathematics constructively, the classically minded referees would look at the theorems, and conclude that they already knew them. They were quite hesitant to accept constructive proofs of known classical results, whether or not constructive proofs were previously available. They did not see this aspect as important. Logicians as referees had a different problem; the papers were mathematics, not logic, and logicians were reluctant to pass judgment on them. Nowadays, with the interest in computational mathematics, things might be different. Bishop said he ceased to take students because of these problems. In the 1970's he said he found it very pleasant and somewhat amusing that it was the logicians and computer scientists who were in the forefront of those who not only appreciated his work, but studied it and were inspired by it -- such well known figures as Harvey Friedman, John Myhill, and Robert Constable.

When Bishop was invited to speak to the AMS 1982 Summer Institute on Recursion Theory, he replied that the aggravation caused by the lecture tour a decade earlier had contributed to a heart attack, and that he was not willing to take a chance on further aggravation. The organizers assured him that he would have a very receptive audience, but he still declined, on the grounds that he did not like to travel anyway.

[M, N] We visited Bishop in 1981 in La Jolla. (Metakides came over from Greece for a month; Nerode was there for six months.) We were primarily interested in adapting his Brouwerian counterexample insights (which show the limits of constructive functional analysis) to recursive functional analysis to show the limits of recursive functional analysis. We told him that with appropriate definitions it was very easy to read every theorem and proof in his book as true in recursive functional analysis. We asked whether he had constructed Brouwerian counterexamples to the obvious possible strengthenings of the theorems in his book. He said that of course he had constructed counterexamples to all obvious strengthenings in every case, otherwise he would not have been doing his job of determining how far constructive functional analysis goes. He said he had suppressed these counterexamples for two reasons. One reason was that he regarded the positive affirmative theorems as the actual mathematics, and the counterexamples as merely showing necessary limits. The other was that he knew there was resistance to understanding the Brouwerian counterexamples, and he did not wish to distract the mathematical reader. We told him that many readers were puzzled by the exact hypotheses of the constructive theorems of the book, since these were absent in the classical theorems. We expressed the view that those trying to use the book would have a much easier time and gain much more confidence in what they were doing if they knew how to determine for themselves exactly why each hypothesis was there. We had read papers by others following up on his book, and it was quite clear that their authors, unlike Bishop, did not possess the knack of constructing with ease the Brouwerian counterexamples to possible strengthenings. We think that Bishop unintentionally rubbed out the record of how positive affirmative mathematics is actually assembled, since such assembly is not as easy for others as it was for him.

It was indeed not only easy, but also natural to him. We asked him if he had been a late convert to constructive methods. This we were curious about, since he was a very well established classical functional analyst for many years before doing his book. He said that the fact of the matter was that he thought by nature constructively from the beginning of his mathematical life, and that the book was a natural outgrowth of his mathematical temperament. (In this he resembles Brouwer, who in later life regarded his sojourn in classical mathematics as a temporary deflection from his natural path.)

From our point of view, Brouwer had seen how to formulate 19th century finite dimensional algebra and analysis from a constructive point of view, while Bishop showed how to extend the constructive point of view naturally to 20th century infinite dimensional functional analysis. We asked him how much he had been influenced by Brouwer's writings. He said that he had been influenced by Weyl's book, and had looked briefly at Brouwer, but had avoided detailed study of Brouwer for fear of being led away from his own natural lines of development.

Bishop had always had an exceptional reputation, among those who knew him, for speed and concentration. He said of the effect of Time on his abilities, "When I look at my mathematical speed now I don't find it lacking in comparison to either mine as a young man or to that of young men now. The difference lies elsewhere. When I was young, I attacked mathematical problems with abandon. Now I think to myself: If I start this project, how many sleepless nights will it cost me? It's the loss of abandon that makes the difference, not the loss of ability."

We asked him whether he had thought out the applicability of his ideas to the rest of mathematics beyond his book. He said that for conventional analysis and algebra it would be by now a fairly routine activity for him to sift out the constructive content, since this amounted to making the constructivised hypotheses sufficiently strong. He said he had made several attempts at a constructive theory of ordinals, but that this still presented him with problems.

We asked him whether he was interested in the computer algorithm implications of constructive methods. He answered that that was not one of his motivations originally, and that the constructive proofs he had offered in his book were not sharp enough for this purpose.

We asked him if he had a principal current project in the area of constructive mathematics, or whether he was leaving it for others. He said that he had spent a lot of mental time reconstructing his book so as to give constructions throughout which are finitely graspable from the data supplied, in the sense that determinants or Newton's method are graspable. We could understand this position well, since our own subject of recursion theory was constructed without regard to sharpness of bounds on time and space for computation.

There is a real correspondence between insights into Brouwerian counterexamples to possible assertions in constructive mathematics, and insights needed to determine the recursive content of these same assertions. Bishop was the expert on the Brouwerian counterexamples. So we translated

each recursion theoretic problem we were interested in into language under-
standable to him in constructive mathematics. He was exceedingly helpful on
every occasion, and always had interesting new insights. He said that he had
to admit that the questions which arose were interesting and worthwhile for
constructive mathematics, whatever their origin in classical mathematics.
The theorem proven in the accompanying paper by Metakides, Nerode, and
Shore is one consequence of those conversations. There are also other such
theorems which arose from those discussions, which will be explored in future
papers.

[C]   Over the past decade, my colleagues and I at Cornell have designed
and implemented a programming language based on constructive mathematics.
This achievement was largely inspired by Bishop's writings, lectures, and
encouragement. It all started in 1970 when I heard a lecture by Richard
Waldinger on the problem of automatic program synthesis from very high level
problem specification. I thought that a formal system of constructive mathematics would serve superbly both as a high level specification language as
Waldinger wanted, but even better as a very high level programming language
itself.

Bishop's book convinced me that this effort was more worthwhile than I
had imagined, and it considerably broadened my view of the project. When he
visited us at Cornell, I learned an attitude toward computation that changed my
whole conception of computing theory. I met Bishop again in 1974 and in con-
versation with him further refined my plan for a programming language. In
1977 a Cornell graduate student, Joseph L. Bates, began to study the role of
such a language in programming methodology. We started working together,
and by 1983 we had designed and built a large computing system which would
execute constructive proofs. We had also by then designed a language which
we believed to be adequate to formalize all of Bishop's book as well as the
book of Aho, Hopcroft, and Ullman on the design of algorithms. Shortly after
we had executed our first constructive proof, I wrote to Bishop informing him
of what I took to be an historic event. I told him how much his writings and
his encouragement had meant to us on the long road to this accomplishment.
I was crushed to receive my letter back unopened, marked "recipient
deceased."

References

Metakides, G. and Nerode, A., The Introduction of Non-recursive Methods in Mathematics, The L. E. J. Brouwer Centenary Symposium, North-Holland, 1982, pp. 319-336.

Metakides, G., Nerode, A., and Shore, R. S., Recursive Limits on the Hahn-Banach Theorem, this volume.

Constable, R. L., "Constructive Mathematics and Automatic Theorem Writers," Proc. of IFIP Congress, Ljubljana, 1971, pp. 229-233.

Bates, J. L., and R. L. Constable, "Proofs as Programs," Dept. of Computer Science Technical Report TR 83-530, Cornell University, Ithaca, New York, 1982 (also to appear in TOPLAS).

Constable, R. L., and Bates, J. L., "The Nearly Ultimate PRL," Dept. of Computer Science Technical Report TR 83-551, Cornell University, April, 1983.

Contemporary Mathematics
Volume **39**, 1985

RECURSIVE LIMITS ON THE HAHN-BANACH THEOREM

G. Metakides, A. Nerode and R. A. Shore

POSITIVE RESULTS

In his book <u>Constructive Analysis</u>, Bishop gives a

<u>Constructive Hahn-Banach Theorem:</u> (Theorem 4, p. 263). Let λ be a non-

zero linear functional on a linear subset V of a separable normed linear

space B, whose null space $N(\lambda)$ is a located subset of B. Then for each

$\epsilon > 0$ there exists a normable linear functional ν on B with $\nu(x) = \lambda(x)$ for

all x in V, and $\|\nu\| < \|\lambda\| + \epsilon$.

For those unfamiliar with his constructive vocabulary, some explana-

tions are in order. The operations of scalar multiplication, vector addition,

and norm for such a space have to be constructively computable, as does the

functional λ. But what does "located" mean, and how about "normable"? A

located subset $N(\lambda)$ is one such that one can constructively compute the dis-

tance $d(x)$ of any point x of B from $N(\lambda)$; for the kernel $N(\lambda)$, this turns

out to be constructively equivalent to assuming one can constructively compute

the norm of λ. A normable functional λ is one whose norm can be construc-

tively computed. The content of Bishop's <u>constructive</u> Hahn-Banach theorem

is that given ϵ, one can compute ν and its norm from constructive descrip-

tions of λ, the norm of λ, vector addition in B, scalar multiplication in B,

and the norm function $\| \ \|$ in B. (The <u>classical</u> Hahn-Banach theorem does

not have an ϵ in it. There is an explanation of the necessity of some restric-

tion below.) In order to state the recursive function version of the correspond-

ing Hahn-Banach theorem it is necessary to sketch the definitions appropriate

to recursive analysis. What does it mean to compute with a real number α?

It means to compute from approximations to α. Let $\{q_0, \ldots, q_n \ldots\}$ be a

standard effective enumeration of the rational numbers. A description of a

real number α is a pair (f, g) of functions of integers such that $|q_{g(n)} - q_{g(m)}|$

$< 1/k$ for all $m, n > f(k)$, and $\{q_{g(n)}\}$ converges to α. (This says that

$\{q_{g(n)}\}$ is Cauchy and $f(n)$ is its "modulus of convergence" function.) A real number α is recursive if it has a description (f, g) with f, g recursive functions. A code for a recursive real number is a pair of numbers (e_1, e_2), e_1, e_2 being Gödel numbers of procedures for computing f and g respectively. A recursively presented metric space consists of a metric space $(X, \delta(x, y))$ and an enumeration $\{x_0, \ldots, x_n \ldots\}$ of a countable dense subset such that certain standard metric space constructions are recursive. The main one is the existence of a uniform recursive procedure which <u>computes</u> the code of recursive real number $\delta(x_i, x_j)$ from i, j. A description of an arbitrary point x of the completion consists of a pair (f, g) of functions of natural numbers such that $\delta(x_{g(n)}, x_{g(m)}) \langle 1/k$ for all $m, n \rangle f(k)$ and $\{x_{g(n)}\}$ converges to x. Such an x is called recursive if f, g can be chosen recursive. Codes of recursive points x are defined as before as pairs of Gödel numbers (e_1, e_2) of procedures for computing f, g respectively. Finally, we need the notion of a recursively continuous function F from one recursively presented metric space S to another such T. A very easy but not very elegant definition envisages an ordinary Turing machine with an infinite tape on which we record descriptions of arguments x of F, and then run the machine to compute descriptions of values $y = F(x)$. Think of the squares on the tape mod 5. If x has description (f, g), the 5 <u>n</u>th square is devoted to recording a pair of integers $(k, f(k))$, the 5n+1 <u>th</u> square to pairs $(\ell, g(\ell))$ so that the whole description of x (and nothing else) is recorded on these squares. The 5n+2 <u>th</u> squares are devoted to computation when the Turing machine is turned on. All the contents of the 5n+3 <u>th</u> squares and the 5n+4 <u>th</u> squares constitute respectively all $(k', f'(k'))$ and all $(\ell', g'(\ell'))$ for some description (f', g') of $y = F(x)$. More succinctly, $F: S \to T$ is recursively continuous if there is a Turing machine such that whenever a description of an argument x is input, a description of value $y = F(x)$ is output. The Turing machines have their usual standard effective enumeration, which thus automatically has a place for each Turing procedure for computing each recursive continuous function on S to T. We let φ_n be the function described by the n <u>th</u> Turing machine.

A recursively presented Banach space consists of a recursively presented complete metric space with recursively continuous operations of scalar multiplication, vector addition and norm which make the space into a Banach space. A recursively located subset is one such that the function $d(x)$ (distance to the subset) is a recursively continuous function. For further details

on definitions of recursive analysis, see Metakides-Nerode [2] or Pour-El-
Richards [3].

Recursive Version of the Hahn-Banach Theorem (cf. Metakides-Nerode [2]).
Let λ be a non-zero recursive continuous linear functional on a linear subset
of a recursive Banach space B whose null space $N(\lambda)$ is a recursively located subset of B. Then for any $\epsilon > 0$, there is a recursively continuous
linear functional ν on B with recursive norm such that $\nu(x) = \lambda(x)$ for all
x in V, and $\|\nu\| \leq \|\lambda\| + \epsilon N(\lambda)$ is recursively located precisely when λ
has a recursive norm.)   Like all the proofs in Bishop's book, the proof of
the constructive Hahn-Banach theorem given there using convex sets can be
read in the sense of recursive analysis. It gives this result. Metakides and
Nerode [2] adapt the other common proof, proceeding by extension of ν from
a finite dimensional subspace S to a new recursive x in B not in S. Recall that the new $\nu(x)$ can (and must) be any number in the interval

$$\sup\nolimits_{u \in S}[\|-u + x\| - \nu(u)] \leq y \leq \inf\nolimits_{v \in S}[\|v + x\| - \nu(v)].$$

This inequality, in the case of recursively presented S and recursively continuous ν on S has a real number as lower endpoint whose lower Dedekind
cut is recursively enumerable, and has a real number as right endpoint with
an upper Dedkind cut which is recursively enumerable. In case these coin-
cide, the common value is the only possible $\nu(x)$, and is a recursive real and
defines the desired extension. In case the endpoints are distinct, any recursive real in the interval is an appropriate $\nu(x)$. So proceeding a finite number
of dimensions is always possible, a recursive value of $\nu(x)$ being chosen for
each new recursive x not in the previous domain S.

Corollary: The recursive Hahn-Banach theorem holds for finite dimensional
recursively presented Banach spaces without the ϵ.

NEGATIVE RESULTS

        What about dropping the ϵ in Bishop's constructive formulation?
Bishop gave a two dimensional counterexample in the Brouwerian style. His
counterexample is not a single space and functional, but rather a family of
such. Let $\mathbb{R}^2$ be the plane, let $\lambda(x, 0) = x$ be defined on the one dimensional
subspace $\mathbb{R}^1$ which is the x-axis. We reformulate his counterexample in our
classical terms as asserting that for a certain family of norms on $\mathbb{R}^2$, there
is no function assigning to each norm $\| \ \|$ of the family a ν extending λ to

$\mathbb{R}^2$, such that λ and ν have the same norm and ν is a continuous function of $\| \ \|$. (Here the norms and linear functionals are given the topology induced by looking at them as functions on the Euclidean $\mathbb{R}^2$ unit sphere under the uniform topology.) Here are three figures.

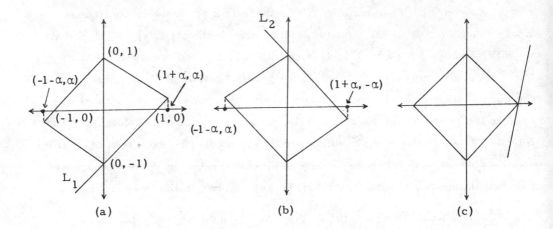

FIG. 1

Let $\| \ \|_1$, $\| \ \|_2$, $\| \ \|_0$ be the norms with respective unit spheres the parallelograms with vertices

$$\{(0, 1), \ (1 + \alpha, \ \alpha), \ (0, -1), \ (-1 - \alpha, \ -\alpha)\},$$

$$\{(0, 1), \ (1 + \alpha, -\alpha), \ (0, -1), \ (-1 - \alpha, \ \alpha)\},$$

$$\{(0, 1), \ (1, 0), \ (0, -1), \ (-1, 0)\}.$$

Suppose the extension $\nu(\| \ \|)$ of λ were a continuous function of $\| \ \|$ and always had the same norm 1 as λ. For sufficiently small α, positive or negative, norms $\| \ \|_1$, $\| \ \|_2$ can be made arbitrarily close to $\| \ \|_0$. But $\nu(\| \ \|_1)$, $\nu(\| \ \|_2)$ are unique extensions, the former maps the line L_1 connecting $(0, -1)$ and $(1, 0)$ to 1, the latter maps the line L_2 connecting $(0, 1)$ and $(1, 0)$ to 1. $(\nu(\| \ \|_0)$ can have a range of values, any extension of λ such that the line mapped to 1 is a line through $(1, 0)$ which does not cut through the parallelogram). Were $\nu(\| \ \|)$ continuous, as α approaches 0 we would get $L_1 = L_2$, contrary to fact.

Since the recursive Hahn-Banach theorem holds in the finite dimensional case with $\epsilon = 0$, there is no two dimensional recursive analogue of the
Bishop example. A recursively presented Banach Space counterexample must
be infinite dimensional. The textbook separable Banach spaces all are recursively presented, but their unit spheres are too smooth to give such an example. What has to be constructed is an infinite dimensional recursively
presented Banach space with an especially prickly unit sphere.

Theorem: There exists a recursively presented Banach space B and a recursive linear functional λ defined on a recursively presented linear subspace V of B such that: 1) λ has a recursively located kernel $N(\lambda)$ (hence
a recursive norm), 2) there is no recursively continuous linear functional ν
on all of B extending λ with the same norm as λ.

Proof: We define a sequence $\{B_i\}$ of two dimensional Banach spaces by
giving an effective definition of an approximation at stage s to the norm
$\| \; \|_i$ on B_i. The underlying vector space for each B_i is $\mathbb{R}^2$. Once this is
done, the Banach space B will consist of all infinite sequences $\{\langle x_i, y_i \rangle\}$
such that all $\langle x_i, y_i \rangle$ are in B_i and such that $\sum_{i=0}^{\infty} \| \langle x_i, y_i \rangle \|_i < \infty$. It will
be convenient to identify each member $\langle x, y \rangle$ of B_i with the element of B
whose ith pair is $\langle x, y \rangle$ and whose other pairs are all $\langle 0, 0 \rangle$. Then V will
consist of those $\{\langle x_i, y_i \rangle\}$ in B such that all $y_i = 0$, and $\lambda(\{\langle x_i, y_i \rangle\}) = \sum_{i=1}^{\infty} x_i$. These definitions will ensure that $\| \langle 1, 0 \rangle \|_i = 1$ and $\| \lambda \| = 1$.

The candidate extensions of λ to all of B are the Turing machine
defined potential recursively continuous mappings on B to $\mathbb{R}$. We let φ_i be
given by the ith Turing machine in the usual enumeration. The idea is to
guarantee that no candidate φ_i for an extension of λ to all of B has a re-
striction $\varphi_i | B_i$ which is a norm preserving extension of the restriction
$\lambda | B_i$. Any extension φ of λ to all of B_i is determined by the value
$\varphi(\langle 0, 1 \rangle)$ since, by linearity, $\varphi(\langle x, y \rangle) = x + \varphi(\langle 0, 1 \rangle) y$.

Consider now the line through $(1, 0)$ on which φ has the value 1. If
this line passes inside the unit sphere of B_i then $\| \varphi \|_i > 1$. Our plan is
therefore to define the norm $\| \; \|_i$ so that for $\varphi = \varphi | B_i$ this line does pass
inside the unit sphere of B_i.

Classically we can easily construct such a norm by choosing one of the
two choices described in figures 2a and 2b below, depending on the value of
$\varphi(\langle 0, 1 \rangle)$. Constructively we must specify a procedure for approximating the

norm based on an approximation to $\varphi(\langle 0, 1 \rangle)$. By being uniform in i for $\varphi = \varphi_i$, such a procedure will yield a recursive norm on B.

Our approximation to $\| \ \|_i$ is given by approximations to the unit sphere of B_i . At stage s, the unit sphere is guaranteed to lie in the shaded area in figure 2c, unless the given approximation to $\varphi(\langle 0, 1 \rangle)$ has been bounded away from +1 or -1. In the latter two cases the unit sphere is defined as in figures 2a or 2b for the least s at which $\varphi(\langle 0, 1 \rangle)$ was found to be bounded away from +1 or -1 respectively.

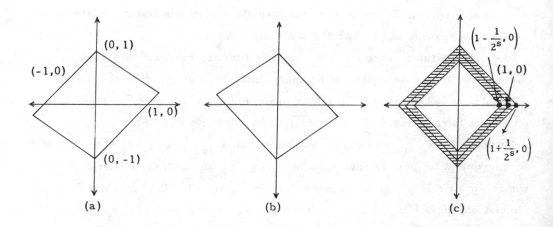

(a)   (b)   (c)

FIG. 2

If $\varphi_i(\langle 0, 1 \rangle)$ is defined, one can compute the unit sphere in B_i as one of the two possibilities in figures 2a and 2b and so prove that $\| \varphi_i \|_i > 1$. In this case φ_i cannot be a norm-preserving extension of λ. The key point, however, is that we have a properly defined norm on B_i regardless of what happens in the computation of $\varphi_i(\langle 0, 1 \rangle)$. As the approximations to $\| \ \|_i$ are done uniformly in i we can compute $\| \ \|$ on the obvious countable dense subset of B given by rational points which are zero on all but a finite number of coordinates. This is all that is needed for obtaining a recursive presentation of B.

Here is another Bishop example about located sets for which there is a connected recursion theoretic theorem. Consider the function $d(x)$ which assigns to a point x its distance from the plane spanned by fixed lines L_1 ,

L_2. The same kind of discontinuity is present as above (look at what happens as L_1 and L_2 become collinear). An argument like that above shows

<u>Theorem</u>: There exists a recursively presented Banach space B and recursively located subspaces X and Z such that neither $X \cap Z$ nor $X \oplus Z$ are recursively located.

Bishop did not accept a closed notion of constructive function. He said that recursive mathematics and constructive mathematics are such that neither contains the other. The use of the enumeration of all Turing machines and the general notion of a partial recursive function implicit in the full argument above would not be acceptable to Bishop. But if $\{\varphi_i\}$ is a constructive list of constructive candidate functionals, then the B constructed is a constructive Banach space and the λ constructed is a constructive normable functional in the sense of his book, and the assumption that some one of these φ_i is the desired extension to the whole of B and has the same norm as λ leads to an absurdity.

REFERENCES

1.   Bishop, E. Foundations of Constructive Analysis, McGraw Hill, New York, 1967.

2.   Metakides, G. and Nerode, A. The introduction of non-recursive methods into mathematics, L. E. J. Brouwer Centenary Symposium Volume, North Holland, 1982, pp. 319-336.

3.   Pour-El, M. and Richards, I. Non-computability in analysis and physics: a complete determination of the class of non-computable linear operators, Advances in Mathematics 48, pp. 44-74.